Physics of Satellite Surface Charging

Physics of Satellite Surface Charging

Causes, Effects, and Applications

Shu T. Lai and Rezy Pradipta

CRC Press is an imprint of the
Taylor & Francis Group, an informa business

First edition published [2022]
by CRC Press
6000 Broken Sound Parkway NW, Suite 300, Boca Raton, FL 33487-2742

and by CRC Press
2 Park Square, Milton Park, Abingdon, Oxon, OX14 4RN

CRC Press is an imprint of Taylor & Francis Group, LLC

Library of Congress Cataloging-in-Publication Data
Names: Lai, Shu T., author. | Pradipta, Rezy, author.
Title: The physics of satellite charging : causes, effects, and applications / Shu T. Lai, Rezy Pradipta.
Description: First edition. | Boca Raton : CRC Press, 2022. | Includes bibliographical references and index.
Identifiers: LCCN 2021041479 | ISBN 9780367224745 (hardback) | ISBN 9781032199214 (paperback) | ISBN 9780429275043 (ebook)
Subjects: LCSH: Space vehicles--Electrostatic charging. | Space plasmas.
Classification: LCC TL1492 .L36 2022 | DDC 629.47--dc23/eng/20220106
LC record available at https://lccn.loc.gov/2021041479

ISBN: 978-0-367-22474-5 (hbk)
ISBN: 978-1-032-19921-4 (pbk)
ISBN: 978-0-429-27504-3 (ebk)

DOI: 10.1201/9780429275043

Typeset in Minion
by MPS Limited, Dehradun

Contents

Foreword

Spacecraft charging received its first important attention when S.E. DeForest [*J. Geophys. Res.*, Vol. 77, pp. 651–659, 1972] observed the ATS-5 spacecraft charged to negative potentials of several kilovolts. The field of spacecraft charging began to stand out when the first four Spacecraft Charging Conferences were held at the US Air Force Academy in the nineteen seventies and eighties. Since then, many journal papers, reviews, and some textbooks have been published. It is now a subject being taught at some universities and colleges.

Two useful textbooks on this subject have appeared in the past three decades. They are Hastings and Garrett, Spacecraft Environment Interactions, Cambridge University Press, [1996], and Lai, Fundamentals of Spacecraft Charging, Princeton University Press, [2012].

This book arose from a request by the publisher, Taylor and Francis, that we expand an article by Lai and Cahoy, entitled "Spacecraft Charging" (in Encyclopedia of Plasma Technology, by Taylor and Francis, 2017) into a book. Attempting to expand this article into a book, we took advantage of our many teaching seminars presented at MIT Space Propulsion Laboratory and at Boston College. It is an honor and privilege to present teaching seminars at MIT and at Boston College. We thank Manuel Martinez-Sanchez, Paulo Lozano, and Patricia Doherty of these institutions.

For the benefit of those reviewers who judge a book by what is absent from it, let us admit that we have omitted laboratory experiments, numerical computation and simulation methods, space instrumentation and engineering, deep dielectric charging, triple-root jumps, alternative to critical temperature, surface material properties and database, discharge initiation, discharge propagations, and more. This book is on the basic physics of spacecraft charging.

The main theme of this book is not to expound, but to try to explain. Einstein was quoted to say "You do not really understand something unless you can explain it to your grandmother!" [L. Maiani, *Euro Phys J.*, doi:10.1140/epjh/e2017-80040-9, 2017]. We thank all the colleagues and students who asked so many good questions. We hope that this book will help, motivate, and interest the readers.

Biography

SHU T. LAI earned his Ph.D. and M.A. from Brandeis University and his B.Sc. from the University of Hong Kong. He earned his Certificate of Special Studies in Administration and Management from Harvard University. He did research at AFRL. He is currently affiliated with the Space Propulsion Laboratory, Massachusetts Institute of Technology, and the Institute of Scientific Research, Boston College. A recognized leader in spacecraft interactions with space plasmas, he has written more than one hundred publications and owns three patents. He is a Fellow of the Institute of Electrical Engineers (IEEE), a Fellow of the Institute of Physics, and a Fellow of the Royal Astronomical Society. He has served as the Chair of the AIAA Atmospheric and Space Environments Technical Committee and the Chair of the AIAA Atmospheric and Space Environments Standards Committee. He is now serving as a Senior Editor of IEEE Transactions on Plasma Science.

REZY PRADIPTA earned his Ph.D. and S.M. in Nuclear Science and Engineering and his S.B. in Physics from the Massachusetts Institute of Technology. He did his postdoctoral work at the Institute for Scientific Research, Boston College. He is currently a senior research scientist at the Institute for Scientific Research, Boston College. His research is about space plasma phenomena and their potential impacts on technological systems using observations from multi-diagnostic instruments such as radars and Global Navigation Satellite Systems (GNSS). These space phenomena include ionospheric plasma density irregularities, traveling ionospheric disturbances (TIDs), acoustic-gravity waves (AGWs), and equatorial plasma bubbles (EPBs). In addition to research activities, he also teaches an upper level undergraduate course (Space Weather and Consequences) at the Department of Earth and Environmental Science, Boston College.

CHAPTER 1

Overview

1.1 WHAT IS SPACECRAFT CHARGING?

When a spacecraft has excess electrical charge of one polarity than that of the other polarity, the spacecraft is charged. For a spacecraft made of conducting material, the electrical charges do not stay inside because of their mutual Coulomb repulsion. Rather, the charges are on the surface of the spacecraft. For a spacecraft made of dielectric material, the electrical charges can be both on the surface and also inside. Even with Coulomb repulsion, the electrical charges inside can hardly be pushed to the surface by the repulsion because the electric conductivity of dielectric materials is low.

In summary, for a conducting spacecraft, surface charging can occur. For a dielectric spacecraft, both surface charging and deep dielectric charging can occur.

1.2 WHAT IS SPACECRAFT POTENTIAL?

Potential is relative. A charged spacecraft has an electrostatic potential ϕ relative to that of space plasma surrounding the spacecraft. In the field of spacecraft charging, it is commonly adopted that the space plasma potential is zero. Therefore, the potential ϕ of a spacecraft charged with excess electrons is negative (i.e. $\phi < 0$).

In short, the spacecraft potential floats with respect to that of the ambient plasma.

DOI: 10.1201/9780429275043-1

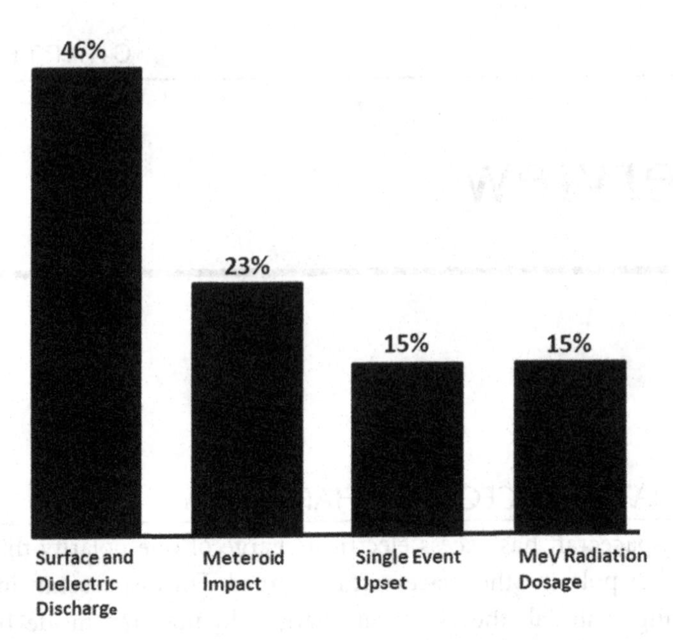

FIGURE 1.1 Causes of spacecraft missions terminated due to the space environment.

1.3 WHY IS SPACECRAFT CHARGING IMPORTANT?

Spacecraft charging may affect the electronic measurements onboard. It may damage electronic instruments. It may also affect the telemetry and even the navigation system. In severe cases, it may terminate the spacecraft mission. Both surface charging and deep dielectric charging are important in causing spacecraft anomalies [Figure 1.1].

1.4 WHERE DOES SPACECRAFT CHARGING OCCUR?

The geosynchronous region is the most important zone for spacecraft charging to occur naturally. In this region, the plasma density varies from about 0.1 to 100 cm^{-3}, and the kinetic energy from multiple kilo-electron volts (keV) to tens of eV. There are many satellites there for communication, surveillance, etc. Natural spacecraft charging is unimportant at low altitudes, where the plasma kinetic energy is low. In the ionosphere, the plasma energy is often about 0.1 eV or less. Spacecraft charging to low voltages is generally unimportant, because low voltage charging is not expected to harm the electronics.

In essence, spacecraft charging is important at geosynchronous altitudes because the ambient plasma kinetic energy, or temperature, may reach keV and above at times, and there are many satellites at such altitudes. There, spacecraft charging can reach kilovolt (kV) negative or, rarely, multiple kV negative.

Spacecraft charging can also occur to a few hundreds of volts negative at auroral altitudes (up to about a few hundred kilometers), where the electrons may become energetic as they flow down the converging magnetic field lines.

Spacecraft charging is not important in the low ionosphere because the ionospheric plasma is usually of energy 0.1 eV or less. The ion density is also relatively high so that if there is any tendency of rising negative voltages, the positive ions nearby are ready to come in and neutralize the negative potential.

1.5 WHEN DOES SPACECRAFT CHARGING OCCUR?

The Sun controls Earth's space weather by sending out solar wind, plasmas, and energetic plasma clouds from time to time. Not all energetic plasma clouds from the Sun are directed at Earth. Solar winds are normally of low energy (up to 10s or 100 eV), but occasionally they are of hundreds of eV. If energetic solar ejecta hit the Earth's magnetosphere, they compress the magnetosphere on the day side and stretch the magnetosphere on the night side. The magnetosphere may be pulled to a long distance beyond the Moon's orbit.

Somewhere (10s or 20s of Earth's radii sometimes) in the elongated magnetosphere, magnetic field merging may occur [Figure 1.2]. The elongated magnetosphere shortens and snaps back toward Earth, thereby squeezing the electrons in the magnetic bottle. The squeezed electrons gain kinetic energy to hundreds or thousands of eV. The phenomenon is called a geomagnetic substorm or storm [*Kivelson and Russell*, 1995; *Lui*, 2000; *Russell*, 2000].

Normally, the magnetosphere without solar disturbances is quiet, with electron energies as low as 10 to 100 eV. In substorms and storms, the electron energy may increase to keV or higher. They are also called hot electrons. Satellites receiving hot electrons may become charged. Hot electrons are more likely at geosynchronous altitudes at about midnight to morning hours.

In summary, hot electrons may charge spacecraft to negative volts. The Sun controls Earth's space weather. The Sun's ejecta may push the dayside of the magnetosphere and pull its night side. Magnetic field reconnection may occur in the elongated magnetosphere on the night side. Following reconnection, the magnetosphere snaps back, thereby

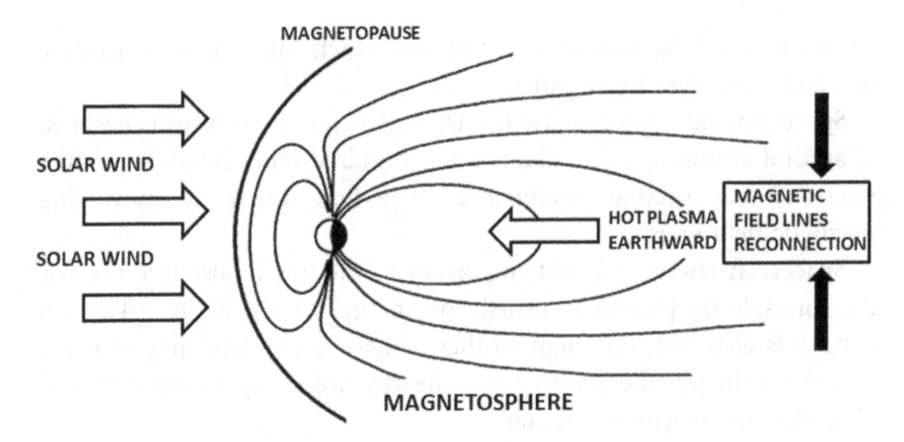

FIGURE 1.2 Push and pull of Earth's magnetic field configuration. The geomagnetic poles and the day–night demarcation line are shown. The magnetopause is pushed by the solar wind. The magnetotail is pulled to long distances of multiple Earth radii. Merging and snap back of the tail cause compression and energization of the charged particles confined by Earth's magnetic lines.

bringing energized, or hot, electrons toward the midnight sector of Earth. The hot electrons drift to the morning hours. Spacecraft charging is more likely during midnight and the morning hours [Figure 1.3].

1.6 ELECTRON AND ION FLUXES

Plasmas consist of electrons and ions. The kinetic energies of the electrons and ions are given respectively by

$$\frac{1}{2} n_e m_e v_e^2 = kT_e \tag{1.1}$$

and

$$\frac{1}{2} n_i m_i v_i^2 = kT_i \tag{1.2}$$

where n denotes the density, m the mass, v the velocity, k Boltzmann's constant, and T the temperature, while the subscripts e and i label electron and ion, respectively. Since nature prefers neutrality, $n_e \approx n_i$. At thermal equilibrium, $T_e \approx T_i$; however, there may be small deviations from perfect balance. As a result, we have the following approximate relation:

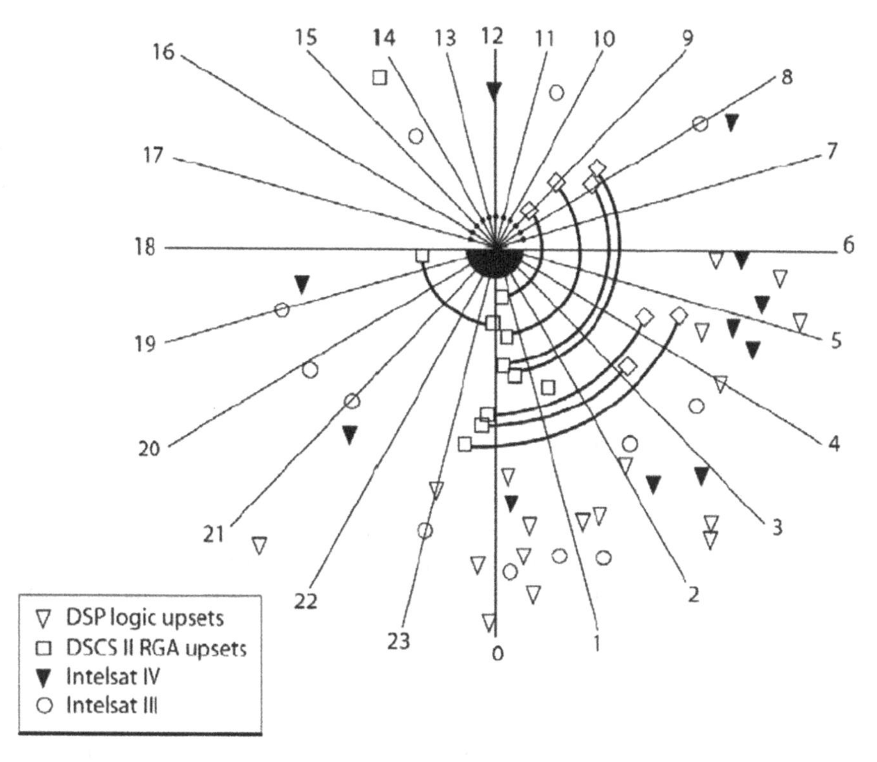

FIGURE 1.3 Spacecraft anomalies as a function of local time. Most of them are in the morning sector. The Sun is in the noon direction.

Source: From Vampola [1987]. Copyright Elsevier. Reprinted with permission.

$$m_e v_e^2 \approx m_i v_i^2 \tag{1.3}$$

Since the electron mass m_e is two orders of magnitude smaller than that m_i of ions, the electron velocity v_e is much faster than the ion velocity v_i. Therefore, the electron flux $qn_e v_e$ is much greater than the ion flux $qn_i v_i$.

$$\frac{1}{2} n_e m_e v_e^2 = kT_e \tag{1.4}$$

Figure 1.4 shows the measured fluxes (Lai and Della-Rose, 2001) of ambient electrons and ions in the geosynchronous environment. One can see that the ambient electron flux is larger than that of ambient ions at all spacecraft potentials by over one to two orders of magnitude. The data at very low potentials are noisy because of reasons that will not

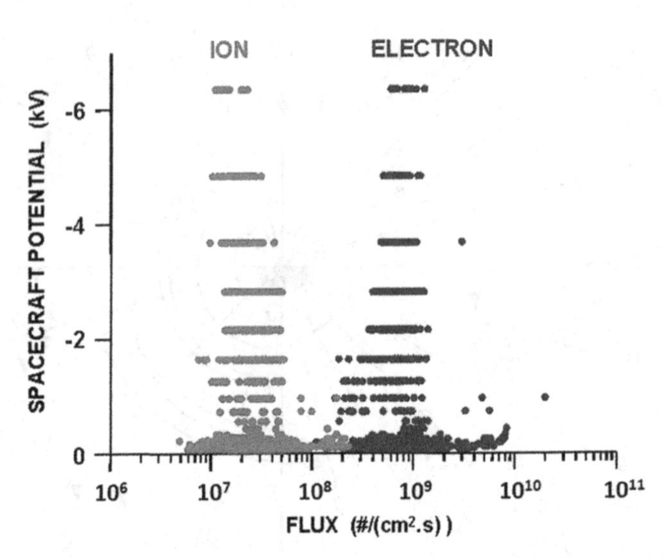

FIGURE 1.4 Electron (blue) and ion (ion) fluxes at geosynchronous altitudes. Data obtained from LANL-1994 Satellite during the eclipse periods, March, 2000.

Source: From Lai and Della-Rose [2001]. Work of US Government. No copyright.

be discussed here. Note that there was a typo in the source by this author, and it has been corrected here.

Physically, the flux inequality, means that an object in space plasmas most likely would intercept more electrons than ions. This result is true not only for space plasmas but also for laboratory plasmas. This simple but important result alone implies that an object placed in laboratory or space plasmas likely charges to a negative potential $\phi(<0)$ because it intercepts more electrons than ions.

In the physics considered so far, there are only two chess pieces in the game: ambient electrons and ions. Later, we will introduce more chess pieces, such as secondary electrons, and the game will be even more interesting.

1.7 GENERAL REFERENCES

For general references on space plasmas, geomagnetic storms, and substorms, see, for example, *Tribble* [1995], *Kivelson and Russell* [1995], *Russell* [2000], *Lui* [2000], *Song, et al.* [2001], *Daglis* [2001], and *Bothmer and Daglis* [2007]. For spacecraft charging, see, for example, *Whipple* [1981], *Tribble* [1995], *Hastings and Garrett* [1997], *Lai* [2011], *Rodgers and Sorensen* [2011], and *Lai and Cahoy* [2017]. For special issues on spacecraft charging in journals, see, for example, *IEEE Transactions on Plasmas, Special Issues* [2006, 2008, 2012, 2013, 2015].

EXERCISES

1. It is good to get an idea of how high the geosynchronous altitude is. Calculate the value of the geosynchronous altitude.
2. The dominant type of ions at geosynchronous altitude is H^+, while the dominant type of ions at ionospheric altitude is O^+. Calculate the ratio of ion mass to electron mass at geosynchronous altitude and at ionospheric altitude. Then, estimate the ratio of electron thermal velocity to ion thermal velocity at geosynchronous altitude and at ionospheric altitude.
3. Suppose the ion mass m_i is 1837 heavier than the electron mass m_e. Let the ion temperature T_i be 5 times higher than the electron temperature T_e. Using the equation $kT = (1/2)mv^2$, calculate the ratio of the ion flux to the electron flux.
4. What is the main reason for spacecraft to charge to negative potentials in space plasmas?
5. Metals (conductors) can undergo surface charging. Can they undergo deep charging? a. Yes. b. No.
6. Dielectrics (insulators) can undergo deep charging. Can they undergo surface charging? a. Yes. b. No.
7. The electric power P of a current I flowing through a resistor equals I^2R, where R is the resistance. The resistor can be an electric light bulb or a computer component. If the current I surges by an amount ΔI, show that the fractional surge $\Delta P/P$ in the power is proportional to $2\Delta I/I$.
8. Spacecraft charging at geosynchronous orbits occurs most likely at about midnight, because
    a. Hot electrons generated in the magnetotail during substorms travel toward the earth.
    b. Solar eclipse in the equatorial plane occurs at about midnight.
    c. Both of the above are true.
9. Natural charging in the ionosphere is insignificant compared with natural charging in the geosynchronous environment. Which of the following is the reason?

a. The ambient electrons in the ionosphere are of low energy (less than 1 eV), except in the auroral region.

b. The ionosphere has high magnetic fields.

c. The ionosphere has lower plasma density.

d. The ionosphere is an ion sphere with no electrons.

10. Which space environment in the following list is most important for natural spacecraft charging?

a. Plasmasphere.

b. Low earth orbits in the ionosphere.

c. Auroral altitudes.

d. Geosynchronous altitudes.

11. When is spacecraft charging likely to occur?

a. During eclipse.

b. During geomagnetic storms or substorms.

c. During periods of high electron temperature.

d. All of the above.

REFERENCES

Bothmer, V., and I.A. Daglis, *Space Weather: Physics and Effects*, Springer-Verlag, Berlin, (2007).

Daglis, I.A., *Space Storms and Space Weather Hazards*, NATO Science Series, Kluwer Academic Publishers, The Netherlands, (2001).

Hastings, D., and H.B. Garrett, *Spacecraft-Environment Interactions*, Cambridge University Press, Cambridge, UK, (1997).

Kivelson, M.G., and C.T. Russell, *Introduction to Space Physics*, Cambridge University Press, Cambridge, UK, (1995).

Lai, S.T. (editor), Spacecraft charging, Vol. 237, *Progress in Astronautics and Aeronautics*, AIAA, Reston, VA, USA, (2011).

Lai, S.T., & K. Cahoy, Spacecraft charging, in *Encyclopedia of Plasmas*, Juda L. Shohet (ed.), Taylor and Francis, New York and London, 10.1081/E-EPLT-120053644, (2017).

Lai, S.T., & D.J. Della-Rose, Spacecraft charging at geosynchronous altitudes: New evidence of existence of critical temperature. *J. Spacecr. Rockets*, Vol. 38, 6, 922–928. 10.2514/2.3764, (2001).

Lai, S.T. (2012). *Fundamentals of Spacecraft Charging*, Princeton University Press, NJ, (2012).
Lui, A.T.Y., Tutorial on geomagnetic storms and substorms, *IEEE Trans. Plasma Sci.*, Vol. 28(6), 1854–1866, 10.1109/27.902214, (2000).
Rodgers, D.J. & J. Sorensen, Internal Charging, Chapter 7 in *Spacecraft Charging*, S.T. Lai (ed.), AIAA Press, Reston, VA, pp. 143–164, (2011).
Russell, C.T., The solar wind interaction with the Earth's magnetosphere: A tutorial, *IEEE Trans. Plasma Sci.*, Vol. 28(6), 1818–1830, 10.1109/27.902211, (2000).
Song, P., H. J. Singer, & G. L. Siscoe, eds., *Space Weather, Geophysical Monograph Series 125*, American Geophysical Union, Washington, DC, (2001).
Special Issue, *IEEE Trans. Plasma Sci.*, Vol. 34(5), 1946–2225, (2006).
Special Issue, *IEEE Trans. Plasma Sci.*, Vol. 36(5), 2218–2482, (2008).
Special Issue, *IEEE Trans. Plasma Sci.*, Vol. 40(2), 138–402, (2012).
Special Issue, *IEEE Trans. Plasma Sci.*, Vol. 41(11), 3302–3577, (2013).
Special Issue, *IEEE Trans. Plasma Sci.*, Vol. 43(9), (2015).
Tribble, A.C. *The Space Environment*, Princeton University Press, NJ, (1995).
Vampola, A.L. Thick dielectric charging on high-altitude spacecraft, *J. Electronics*, Vol. 20, 21–30, (1987).
Whipple, E.C. Potential of surface in space. *Rep. Prog. Phys.*, Vol. 44, 1197–1250, 10.1088/0034-4885/44/11/002, (1981).

CHAPTER 2

Spacecraft Equilibrium Potential

2.1 AN INTRODUCTORY DIALOG

Can a small number of electrons added to a spacecraft surface charge the surface to a high potential? As an exercise, let us imagine putting a negative charge $Q = -60 \times 10^{-9}$ Coulomb on a spacecraft. The potential ϕ at the surface of a conducting sphere of radius $a = 1$ m is given by the Gauss law:

$$\phi = \frac{1}{4\pi\varepsilon_0}\frac{Q}{a} \tag{2.1}$$

Eq (2.1) gives a potential $\phi = -540$ V. Does the result of this exercise imply that objects in space can be easily charged to highly negative potentials? At first thought, the answer seems to be yes because there are so many electrons in space.

However, in nature, a spacecraft would not easily charge to a highly negative voltage even though adding electrons to a spacecraft surface would increase the negative potential of the surface. Two reasons are as follows.

2.1.1 Repelling of Like Charges

Adding electrons to a surface raises the potential (negative volts), but the potential repels the electrons that try to come in later. An incoming

DOI: 10.1201/9780429275043-2

electron of negative charge q with kinetic energy E cannot go through a negative potential ϕ if the kinetic energy E is less than the potential energy $q\phi$. Note that the potential energy, i.e. the product $q\phi$, is positive.

Only those electrons energetic enough to overcome the potential can come in. Therefore, the resultant potential depends very much on the energy distribution of the incoming electrons. Consider a plasma of temperature T in space. Let $I(\phi)$ be the electric current overcoming the spacecraft potential ϕ and $I(0)$ the electric current in the space plasma far away from the spacecraft. It is often a good approximation to adopt the Boltzmann repulsion formula:

$$I(\phi) = I(0)\exp(-q\phi/kT) \tag{2.2}$$

where k is the Boltzmann constant. Note that both $q\phi$ and kT are energies and positive.

2.1.2 Attraction of Opposite Charges

Space is not a vacuum. There are electrons and ions. Most of the ions are O^+ in the ionosphere and H^+ at the magnetosphere. If the spacecraft potential is negative and finite, the positive ions are attracted to the spacecraft surface. Therefore, the resultant potential is determined by the balance of the electron and ion currents coming into and going out of the surface. This is another important property that will be discussed later, in Chapter 3.

2.2 TRANSIENT CHARGING

Suppose the ambient electron current I is constant. It takes time τ to charge a spacecraft of surface capacitance C to a potential ϕ. As an analogy, it takes time τ to fill some tap water of current I to a beaker of cross-sectional area C up to a water level ϕ [Figure 2.1].

$$\tau I = C\phi \tag{2.3}$$

At any time $t < \tau$, the water level $\phi(t)$ is given by

$$\phi(t) = \frac{I}{C}t \tag{2.4}$$

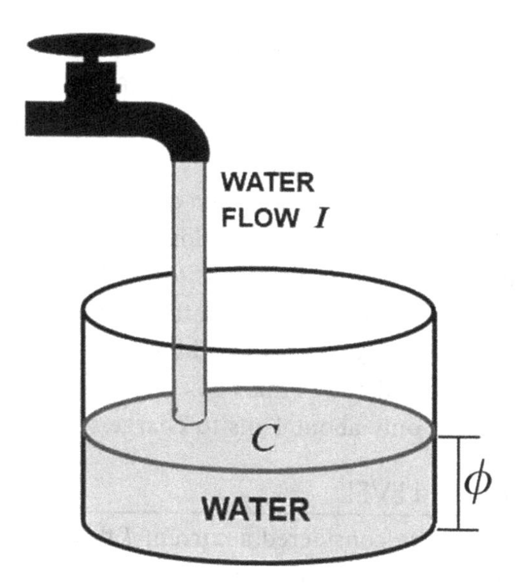

FIGURE 2.1 Concept of capacitance charging. Water current I flows into a container with cross-section C. It takes time τ to fill the beaker of water to a height ϕ.

If one knows the constant current I and the level $\phi(t)$ at time t, one can determine the capacitance C.

$$C = \frac{I}{\phi(t)} t \tag{2.5}$$

If the current I is time dependent, the level $\phi(t)$ is given by

$$\phi(t) = \int_0^\tau dt \frac{I(t)}{C} \tag{2.6}$$

In eq (2.6), we have assumed that the initial $\phi(0) = 0$. If it is not 0, we have to replace eq (2.6) with

$$\phi(t) - \phi(0) = \int_0^\tau dt \frac{I(t)}{C} \tag{2.7}$$

The capacitance C for a spherical surface in space plasmas, such as those at geosynchronous altitudes and beyond, can be approximated as if it were in vacuum. For example, the time τ to charge a spherical spacecraft of 1 m radius to 1 kilovolt (kV) with a flux of 0.5 nA/cm^2 at geosynchronous altitudes is given by eq (2.3), which yields a result of the order of a millisecond (ms), which is fast.

However, if a spacecraft has a higher capacitance, it would take a longer time to charge. For example, thin double layer capacitors or high-density plasmas would increase the capacitance. It is possible to have a hypothetical situation in which a spacecraft with thin capacitor surfaces requires a charge time longer than the orbital period, but such a case is rare and will not be considered in this book. For most practical purposes, we know that it takes only about 1 ms to charge.

2.3 EQUILIBRIUM LEVEL

In the above section, we considered a current I flowing to a spacecraft. The potential $\phi(t)$, in response, increases as long as the current, I, continues flowing to the spacecraft. Suppose there is another current I_i of the opposite electrical charge flowing to the spacecraft. The net charges collected by the spacecraft would be governed by the two currents carrying opposite electrical charges.

As an analogy, think about the water level in a beaker, which is receiving a current I but also leaking away a current I_i. As a result, there will be an equilibrium level. The equilibrium level is governed by the rate of the incoming negative charge and that of the leaking negative charge.

However, unlike water, charges in space plasmas have two signs/polarities. The equilibrium level is governed by the rate of the incoming negative charges and that of the incoming positive charges.

2.4 FLOATING POTENTIAL

It is customary to consider the spacecraft potential as relative to the ambient plasma potential, which is defined as zero (0 V). In this way, the spacecraft potential is regarded as a floating potential, because it can float up or down relative to the constant sea level (0 V) of the ambient plasma [Figure 2.2].

There is a sheath region in which the potential $\phi(r)$ varies gradually from the surface potential ϕ to the ambient plasma potential 0. It is often a good approximation to describe the sheath potential profile $\phi(r)$ in space as [*Whipple, et al.*, 1974]:

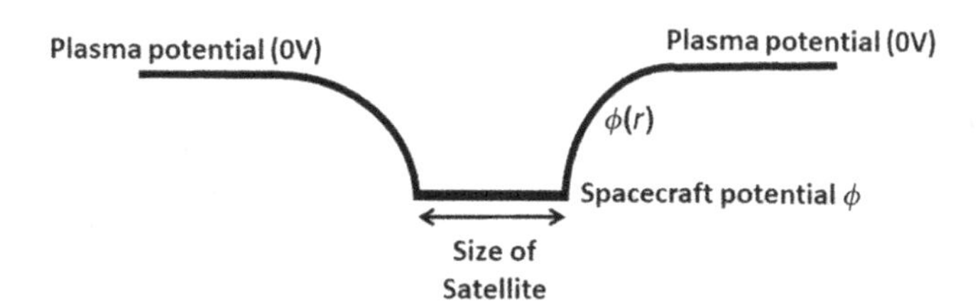

FIGURE 2.2 Concept of floating potential. The ambient plasma potential is defined as zero (0 V). The satellite potential is ϕ at the satellite surface ($r = 0$). There is a sheath in which $\phi(r)$ varies gradually from the surface potential ϕ to the ambient plasma potential 0.

$$\phi(r) = \phi(0)\frac{R}{r + R}\exp\left(-\frac{r}{\lambda_D}\right) \tag{2.8}$$

where R is the radius of the satellite body, r is the radial distance from the body, and λ_D is the Debye radius. For the SCATHA satellite at geosynchronous altitudes, the Debye radius λ_D of the ambient plasma is about 45 m [*Aggson, et al.*, 1983]. At ionospheric altitudes, the Debye radius is much shorter (a few centimeters).

2.5 ELECTRON AND ION ENERGIES IN A SHEATH

Figure 2.3 shows the sheath potential profile $\phi(r)$ in relation to the space plasma potential, which, by convention, is zero. The profile $\phi(r)$ is a function of distance r from the spacecraft surface. The profile shown in Figure 2.3 is attractive for positive ions coming from the right-hand side and repulsive for electrons coming likewise. Asymptotically, the sheath potential $\phi(r)$ approaches that of the space plasma.

2.5.1 Incoming Electrons

For an electron coming toward the spacecraft, the effect of the negative potential is to slow down the electron. During this process, the electron kinetic energy decreases and its potential energy increases, while the total energy E_{total} is constant.

$$E_{\text{Total}} = (1/2)mv^2(r) + q_e\phi(r) \tag{2.9}$$

where m is the electron mass, v the velocity, x the radial distance, $q_e(<0)$ the electron charge and $\phi(<0)$ the potential. The initial energy from infinity is given by

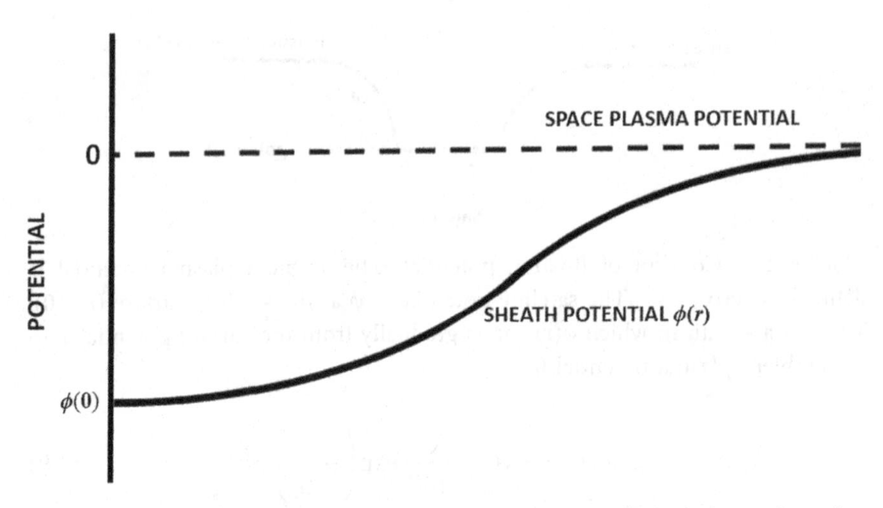

FIGURE 2.3 Sheath potential profile. The satellite surface is located at the y-axis. The y-axis also depicts the scale of the potential. The x-axis depicts the distance r from the surface. The space plasma potential is, by convention, zero. The spacecraft sheath potential $\phi(r)$ located at a distance r from the surface is shown. The potential $\phi(0)$ is the potential at distance $r = 0$ from the surface, and it is the spacecraft surface potential.

$$E_{\text{Total}} = (1/2)mv^2(\infty) + q_e\phi(\infty) \tag{2.10}$$

where $\phi(\infty) = 0$. If the initial velocity is high enough, the electron can reach the spacecraft, but its final kinetic energy will be reduced by $q_e\phi(0)$.

$$(1/2)mv^2(r) = (1/2)mv^2(\infty) - q_e\phi(r) \tag{2.11}$$

When $q_e\phi(r)$ reaches E_{total} at a point r, the velocity $v(r)$ is zero. There, the electron stops at r and will start to move away from the satellite. If one plots the potential profile to depict the uphill travel of the electron from right to left, one can see intuitively that the electron stops at a point r and will start to roll down the hill again [Figure 2.4].

2.5.2 Outgoing Electrons

As an exercise, suppose an electron travels from left to right, starting from $r = 0$. Since an electron has a negative charge, the potential profile accelerates the electron as it travels from the high negative potential. As an analogy, the acceleration from left to right is like a ball rolling downhill. Write down the conservation of energy, and present the result.

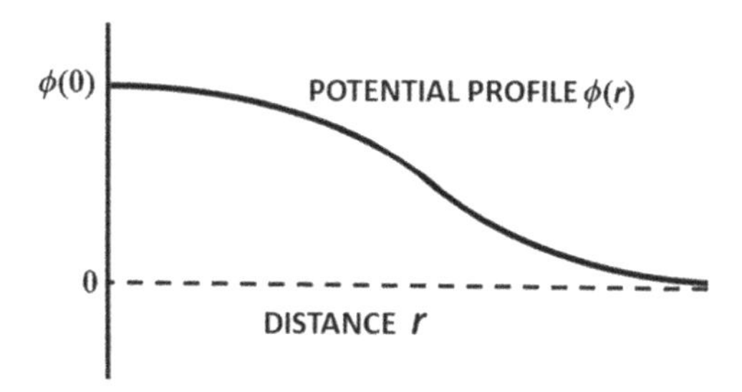

FIGURE 2.4 The potential profile with negative voltage on the upper half plane. An electron traveling from the right side to the left would slow down because of the uphill (negative) voltage for the electron (with negative charge).

2.5.3 Incoming Positive Ions

As a second exercise, write down the total energy for a positive ion coming in from right to left. The ion would travel faster and faster as it approaches the spacecraft, which has a negative potential $\phi(0)$.

2.5.4 Outgoing Ions

As a third exercise, consider a positive ion with kinetic energy E_1 emitted from the spacecraft surface. The initial kinetic energy is $E_1 = (1/2)MV^2(0)$, where M is the ion mass, and V is its velocity. The spacecraft potential ϕ is negative. The total energy is $E_1 + q_i\phi$, where q_i is positive. If the ion energy E_1 is less than the magnitude of the potential energy $q_i\phi$, the total energy is less than 0, which is the potential energy of the ambient plasma.

In Figure 2.5, we have denoted the spacecraft surface potential as ϕ, which is the same as $\phi(0)$ as a function of the radial distance r. As the ion moves along r, its velocity $V(r)$ and the potential energy $\phi(r)$ at distance r satisfy the energy equation.

$$\frac{1}{2}MV^2(r) + q_i\phi(r) = \frac{1}{2}MV^2(0) + q_i\phi(0) \tag{2.12}$$

The velocity $V(r)$ becomes zero at the radial distance r when

$$q_i(\phi(r) - \phi(0)) = \frac{1}{2}MV^2(0) \tag{2.13}$$

where $E_1 = (1/2)MV^2(0)$. The ion stops at r where $V(r) = 0$.

As an exercise, show that for an initial ion kinetic energy $E_2 > 0$, the ion goes to infinity. What is the final ion kinetic energy?

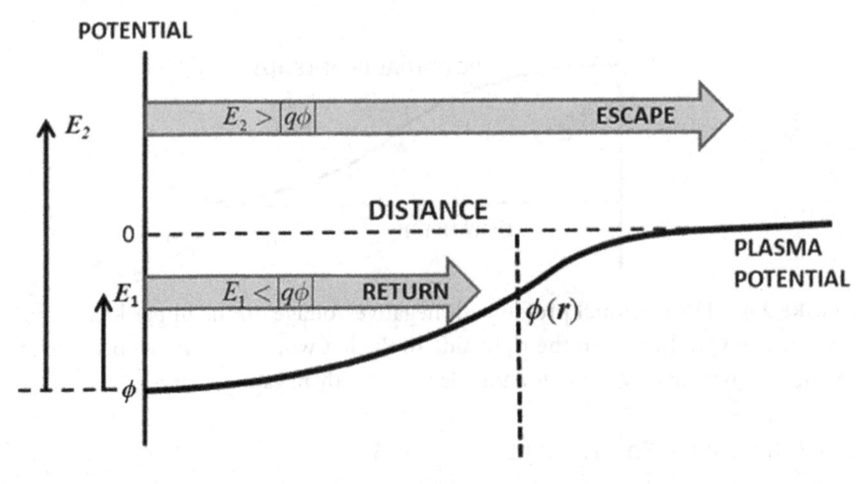

FIGURE 2.5 Energy of a positive ion emitted from the surface. The surface potential $\phi(0)$ is negative, and the space plasma potential is zero. The horizontal axis depicts the distance r from the spacecraft surface.

EXERCISE

1. The time τ for charging a spherical surface to a potential ϕ is proportional to $\phi/(r,J)$, where r is the radius and J the ambient current density. It can be shown that it takes about 2 ms to charge a sphere of $R = 1$ m to $\phi = -1$ kV at the geosynchronous environment. How long would it take to charge a spherical dust particle of radius 10^{-6}m to 2 kV in the same environment?

REFERENCES

Aggson, T.L., B.G. Ledley, A. Egeland, and I. Katz, Probe measurements of dc electric fields, *Eur. Space Agency Publ.* SP-198, 13–17, (1983).

Whipple, E.C., J.M. Warnock, and R.H. Winckler, Effect of satellite potential in direct ion density measurements through the magnetosphere, *J. Geophys. Res.*, Vol. 79, 179, (1974).

CHAPTER 3

Current Balance

3.1 LANGMUIR'S ATTRACTION FORMULA

Consider a charged particle of charge q, mass m, and velocity v, traveling along a straight line without any field of interaction. Let us consider the particle to be a positive ion with its thermal energy kT, where k is Boltzmann's constant and T is temperature.

$$\frac{1}{2}mv^2 = kT \tag{3.1}$$

When the particle gets near a sphere with an attractive field $\phi(r)$, where r is radial distance from the center of the sphere, the particle velocity becomes $v(r)$. Since the total energy (kinetic plus potential) is conserved, the particle energy equation is given by

$$\frac{1}{2}mv^2 = \frac{1}{2}mv^2(r) + q\phi(r) \tag{3.2}$$

where q is the elementary charge, which is positive for positive ions, and $\phi(r)$ is negative for an attractive field by a negatively charged sphere. Since the sign of $q\phi(r)$ is negative, the velocity $v(r)$ increases when the distance r decreases as the particle is approaching the sphere. This is illustrated in Figure 3.1.

The total angular momentum in a central field is conserved and is therefore constant at any point r.

DOI: 10.1201/9780429275043-3

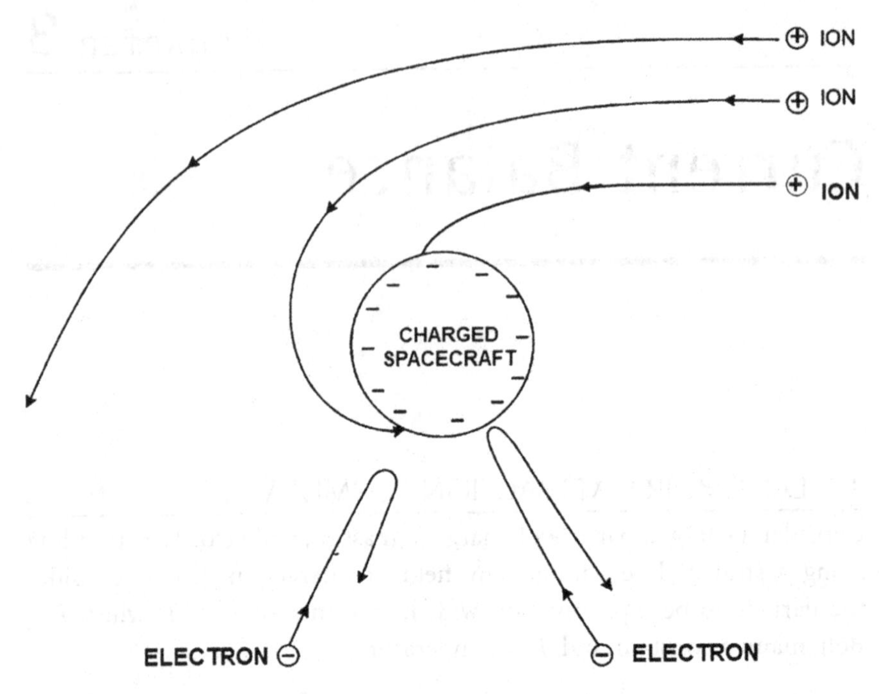

FIGURE 3.1 A schematic diagram of orbit-limited Langmuir attraction of opposite charges. Repulsion of like charges is also shown.

$$mvh = mv(r)r \tag{3.3}$$

where h is the impact parameter, which is the perpendicular distance from the straight line to the sphere center. When the particle touches the sphere of radius a, its velocity $v(a)$ satisfies eqs (3.4 and 3.5).

$$\frac{1}{2}mv^2 = \frac{1}{2}mv^2(a) + q\phi(a) \tag{3.4}$$

and

$$mvh = mv(a)a \tag{3.5}$$

Let us denote ϕ as the potential of the sphere. That is, $\phi = \phi(a)$.

From (3.4), we have

$$v(a) = v\left(1 - \frac{q\phi}{kT}\right)^{1/2} \tag{3.6}$$

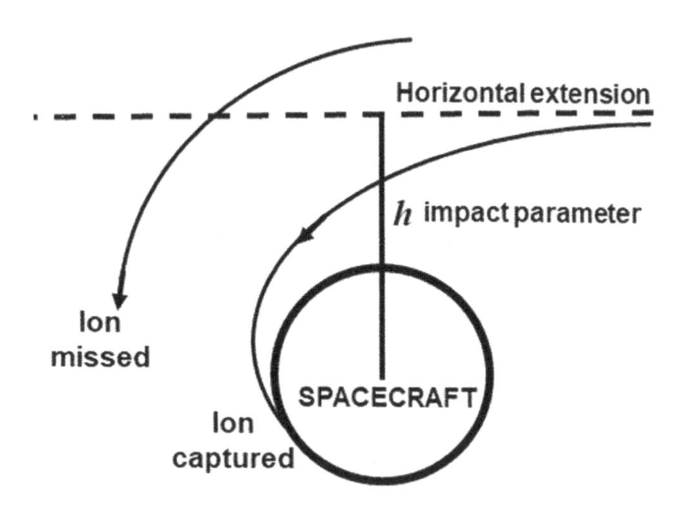

FIGURE 3.2 Concept of impact parameter. The ion travels horizontally from infinity. As it enters the attraction field of the spacecraft, it spirals toward the attraction center. There is a vertical distance from the center to the horizontal extension of the trajectory. There exists a maximum vertical distance h, beyond which the ion trajectories would miss the sphere.

Therefore, the impact parameter h of (3.5) is related to a, v, and ϕ as follows.

$$h = v\left(1 - \frac{q\phi}{kT}\right)^{1/2} a \tag{3.7}$$

It is important to note that, since $q\phi$ is negative in the case of positive ions, $v(a) > v$ in (3.6) and $h > a$ in (3.7). (3.7) is the Mott-Smith and Langmuir attraction formula, or simply Langmuir's attraction formula [Figure 3.2].

3.2 INCOMING CURRENT

For charged particles of velocity v, the flux J is defined as the number of particles crossing a unit area per unit time. That is, $J = qnv$, where n is the charge density. The current I to an area A is given for the flux crossing the area A. That is, $I = qnvA$.

For an uncharged sphere of radius a and zero potential (i.e. $\phi = 0$), the surface area A is $4\pi a$, and the current $I(\phi)$, where $\phi = 0$, received by the sphere is

$$I(0) = 4\pi a^2 qnv \tag{3.8}$$

The significance of the Langmuir attraction formula, (3.7), is explained as follows. If the perpendicular distance d of the extended straight line trajectory of the charged particle equals h, the particle trajectory touches the sphere. If the distance d is less than h, the particle hits the sphere. But, if the distance d exceeds h, the particle misses the sphere.

In 3D, the charged particles are coming from all directions. One can imagine an effective cylindrical tube of radius h from the sphere center. All charged particles with perpendicular distance $d > h$ will miss the sphere, whereas those with $d \leq h$ will reach the sphere. The surface area of the sphere is $4\pi h^2$, which is greater than $4\pi a^2$. The current $I(\phi)$ received by the sphere is

$$I(\phi) = 4\pi h^2 qnv \tag{3.9}$$

Note that $I(\phi)$ is greater than $I(0)$ because the area of the sphere with radius h is greater than that of the uncharged sphere of radius a. In the uncharged case ($\phi = 0$), all particles within a circular cross-section of radius a would be received by the sphere, whereas in the electrically charged case ($\phi \neq 0$), all particles within a circular cross-section of radius h would be received by the sphere. The increase in the incoming current to the sphere is caused by the Coulomb attraction of the potential field ϕ.

Substituting (3.7) into (3.9), we obtain

$$I(\phi) = I(0)\left(1 - \frac{q\phi}{kT}\right) \tag{3.10}$$

In eq (3.10), $I(\phi) > I(0)$ because $q\phi$ is negative (our particles are positive ions). The multiplicative term in the parenthesis of eq (3.10) is called the Langmuir factor, which exceeds unity.

3.3 CURRENT BALANCE OF ELECTRON AND ION CURRENTS

Consider a spherical satellite receiving current I_e of electrons and current I_i of ions. At equilibrium, the potential ϕ is determined by the sum of all currents, which balance each other.

$$I_e(\phi) + I_i(\phi) = 0 \tag{3.11}$$

where the currents, I_e and I_i, are of opposite signs. Substituting (3.2) and eq (3.10) into eq (3.11), we have the current balance equation for the electron and ion currents as follows.

$$I_e(0)\exp\left(-\frac{q_e\phi}{k\,T_e}\right) + I_i(0)\left(1 - \frac{q_i\phi}{k\,T_i}\right) = 0 \tag{3.12}$$

In eq (3.12), the spacecraft potential ϕ is negative, $q_e\phi$ is positive because the electron elementary charge q_e is negative, and $q_i\phi$ is negative because q_i is positive. The electron and the ion currents balance each other.

For an easy understanding of current balance, we can think of the earlier example of water level at equilibrium in a beaker. As another analogy, we can think of a junction of connected garden hoses in which water runs in and out of the junction without inflation or deflation. Or, in electrical engineering terminology, Kirchhoff's law states that the sum of currents equals zero at every junction of an electrical circuit. In other words, the spacecraft behaves as a junction of currents in a circuit in space.

It can also be said that, in spacecraft charging, a spacecraft behaves as a form of Langmuir probe. However, there is some conceptual difference, although the current balance equation is applicable in both cases. In a laboratory setting, one varies the applied potential to a Langmuir probe and measures the resultant currents, whereas in space, the incoming and outgoing currents vary naturally, and one measures the resultant spacecraft potential.

3.4 BALANCE OF MULTIPLE CURRENTS

If there are more currents coming into or going out of a spacecraft or a spacecraft surface, the charging potential ϕ at equilibrium is governed by the current balance of all currents, according to Kirchhoff's law.

$$\sum_n I_n(\phi) = 0 \tag{3.13}$$

where the subscript n labels the various currents. For example, if there is an artificial charged particle beam with current I_B emitted from the spacecraft, the current balance equation is as follows.

$$I_e(0)\exp\left(-\frac{q_e\phi}{k\,T_e}\right) + I_i(0)\left(1 - \frac{q_i\phi}{k\,T_i}\right) + I_B(\phi) = 0 \tag{3.14}$$

In (3.14), the added beam current $I_B(\phi)$ can affect the spacecraft potential ϕ, and vice versa.

As another example, if there are current I_S of secondary electrons, current I_B of backscattered electrons, and current I_{ph} of photoelectrons, the current balance equation for the charging potential ϕ of the spacecraft surface at equilibrium is as follows.

$$I_e(0)\exp\left(-\frac{q_e\phi}{kT_e}\right) + I_i(0)\left(1 - \frac{q_i\phi}{kT_i}\right) + I_B(\phi) + I_S(\phi) + I_B(\phi) + I_{ph}(\phi) = 0 \quad (3.15)$$

EXERCISES

1. Consider taking into account the effect of sheath around a sphere in space plasma, as alluded to in Chapter 2. In Langmuir's attraction formula, what do you think would happen to h if the Debye length λ_D becomes very small in comparison to the size of the sphere? What would happen to h if the Debye length λ_D becomes very large in comparison to the size of the sphere? Would h increase or decrease if λ_D increases?
2. In the orbit-limited regime, the current $I(\phi)$ collected by a Langmuir probe of potential ϕ depends on ϕ and the geometry (sphere, long cylinder, or plane). Sketch the graphs of $I(\phi)/I(0)$ versus $(1 - q\phi/kT)$ starting from $\phi = 0$.
3. In the orbit-limited regime, the ion current $I(\phi)$ attracted to a sphere of potential ϕ is given by

$$I(\phi) = I(0)\left(1 - \frac{q_i\phi}{kT_i}\right)$$

4. Which of the following kinds of ions are more effective in neutralizing a sphere charged a priori to a high negative potential ϕ?

   a. High temperature ions.

   b. Low temperature ions.

   c. Negative ions.

   d. Large Hadron Collider (LHC's) high energy ions.

CHAPTER 4

How to Measure Spacecraft Potential

4.1 ENERGY DISTRIBUTION

Suppose a spacecraft is charged to a negative potential ϕ. The spacecraft surface repels ambient electrons and attracts ambient positive ions. Each attracted ion reaching the spacecraft surface gains an energy $q\phi$, where q is the ion elementary charge. The energy gained is in electron volts (eV).

As an illustration, consider the ambient ions in a Maxwellian distribution $f(E)$, where E is the ion energy ($E > 0$). When the ions are far away from the spacecraft, they do not feel the attractive potential ϕ.

$$f(E) = n\left(\frac{m}{2\pi kT}\right)^{3/2} \exp\left(-\frac{E}{kT}\right) \tag{4.1}$$

where n is the plasma ion density, m the ion mass, k Boltzmann's constant, and T the ion temperature. Strictly speaking, eq (4.1) is a velocity distribution with the velocity v converted to energy E by $(1/2)mv^2 = E$. Measurements of electrons and ions are in terms of energy E, not velocity.

When the ions reach the charged spacecraft surface, each ion gains energy $q\phi$. Therefore, the ion distribution $f_\phi(E)$ measured on the surface is as follows.

DOI: 10.1201/9780429275043-4

$$f_\phi(E) = n\left(\frac{m}{2\pi kT}\right)^{3/2} \exp\left(-\frac{E - q\phi}{kT}\right) \quad (4.2)$$

where $E > q\phi$. We observe from eq (4.1) and eq (4.2) the following properties:

$$f(E) = 0 \quad \text{for } E < 0 \quad (4.3)$$

$$f(0) = f_\phi(q\phi) = n\left(\frac{m}{2\pi kT}\right)^{3/2} \quad (4.4)$$

$$f(0) = f_\phi(q\phi) \quad (4.5)$$

and

$$f_\phi(E) = 0 \quad \text{for } E < q\phi \quad (4.6)$$

Let us take the log of both sides of eq (4.1).

$$\log f(E) = \log f(0) - \frac{E}{kT} \quad (4.7)$$

If we plot $\log f(E)$ as a function of E, we obtain a straight line with a slope of $-1/kT$. Similarly, from eq (4.2), a plot of $\log f_\phi(E)$ versus E is also a straight line with the same slope. Figure 4.1 shows the graphs of $\log[f(E)]$ and $\log[f_\phi(E)]$ as functions of energy E.

There is a gap between $E = 0$ and $E = q\phi$ in the graph of $\log[f_\phi(E)]$. The gap is formed because of the shift of the original distribution function $f(E)$ to $f_\phi(E)$ by an amount $q\phi$. The shift occurs because every attracted particle arriving at the surface has gained an energy $q\phi$. Even a particle originally at rest (energy $E = 0$) has gained an energy $E = q\phi$ when it arrives. The gap indicates the spacecraft potential ϕ.

4.1.1 Repelled Species

For the repelled charged species (electrons), the distribution shifts to the left on the plot. This is because every particle loses an energy $q\phi$ when it arrives at the surface. A part of the distribution shifts to the left of the

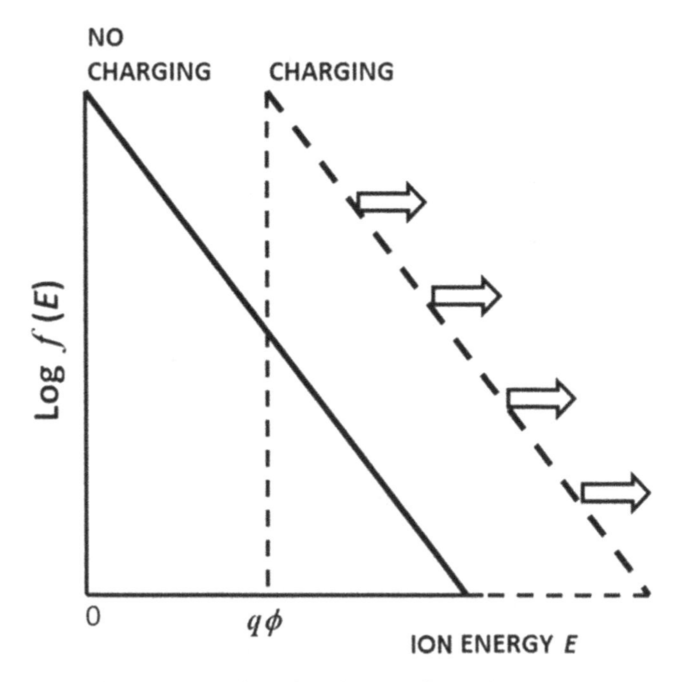

FIGURE 4.1 Shift of a Maxwellian distribution $f(E)$. The energy gap $q\phi$ indicates the spacecraft potential ϕ. The slope, which is $-1/kT$, indicates the temperature T.

y-axis. Since energy cannot be negative, the part shifted to negative is lost. It represents a lost population.

4.1.2 Enhanced Graph

In practice, one way to see the shift clearer is to multiply E to $\log[f_\phi(E)]$. The graph of $E \log[f_\phi(E)]$ plotted as a function E is enhanced. The gap revealed is of the same size.

$$E f_\phi(E) = 0 \quad \text{for } E < q\phi \tag{4.8}$$

4.1.3 Non-Maxwellian Distribution

If the distribution deviates from Maxwellian, the plot of $\log f(E)$ versus E would not be a straight line. Since every particle attracted to the surface would have gained an energy $q\phi$, the gap size is unchanged and can be used to indicate the spacecraft potential [Figure 4.2]. If the deviation from

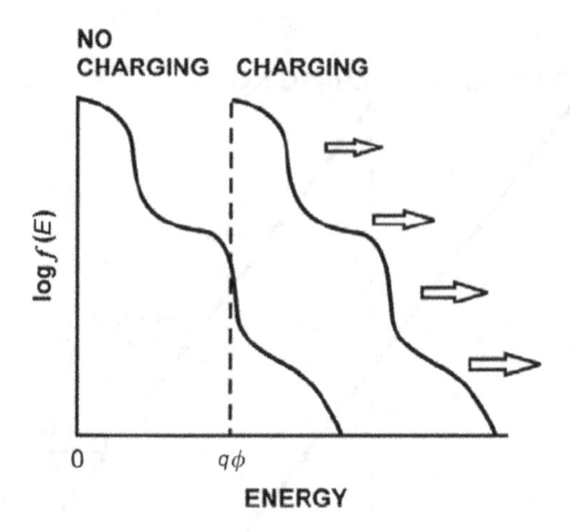

FIGURE 4.2 Shift of a non-Maxwellian distribution.

Maxwellian is too non-uniform or too unsteady, it would be difficult to consider the spacecraft potential.

4.2 INSTRUMENTS FOR MEASURING DISTRIBUTIONS

4.2.1 Retarding Potential Analyzer

Retarding potential analyzer (RPA) has been used for many years for measuring the energy distribution of incoming charged particles. It has a grid plate at the entrance [Figure 4.3]. Suppose a positive potential is applied to it for repelling the incoming ions. Only those more energetic than the applied potential can come in. By varying the repelling potential in increments, one obtains a distribution in discrete points [Figure 4.4]. For more details, see, for example, *Lai and Miller* [2020].

4.2.2 Disadvantages of RPA

Applying small constant increments for a large potential range would take a long time. For approximation, one may use constant increments in log scale, rendering smaller increments for low potentials and longer increments for high potentials. Another disadvantage is the generation of secondary electrons [*Baragiola*, 1993; *Lin and Joy*, 2005] by ion impact on the grids. One can install successive grids to repel the secondary electrons, but it is impossible to repel them all because there is always an innermost grid.

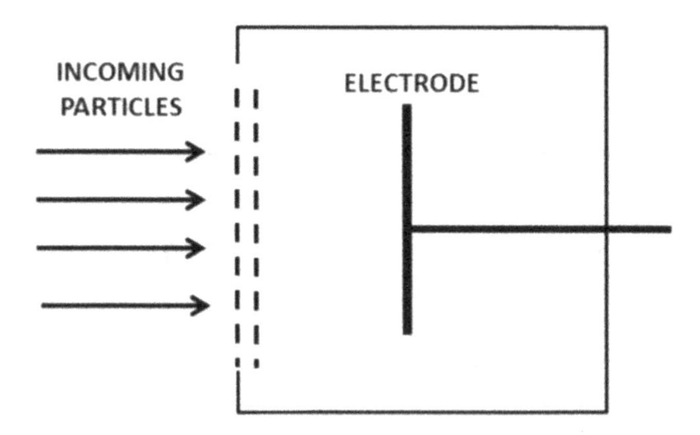

FIGURE 4.3 A schematic diagram of a retarding potential analyzer. The potential applied to the entrance grid repels the charged particles with energies below the grid potential energy.

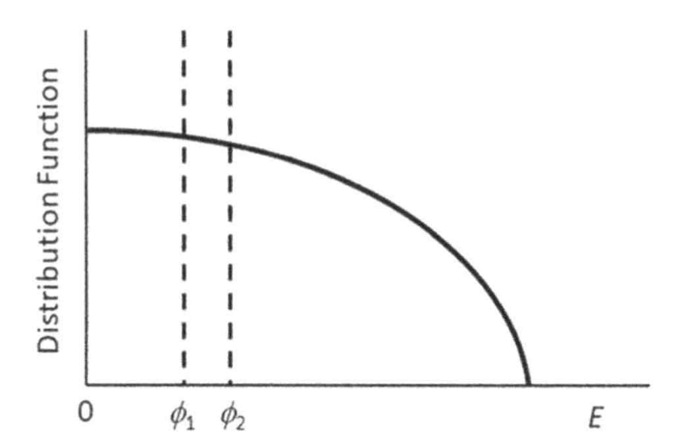

FIGURE 4.4 Measuring a charged particle distribution by varying the grid potential in increments.

4.2.3 Plasma Analyzer

Another popular type of instrument for measuring charged particle distributions is the plasma analyzer. It is also called electrostatic analyzer or "top hat" analyzer [*Vampola*, 1998].

$$\frac{d\mathbf{v}}{dt} = \frac{q}{m}(\mathbf{E} + \mathbf{v} \times \mathbf{B}) \tag{4.9}$$

In (4.9), v is the particle velocity, m the mass, and q the charge. **E** and **B** are the applied electric and magnetic fields, respectively, and t is time. The principle is essentially that of a mass spectrometer, although, in this case, the charged particles have the same mass with an energy distribution.

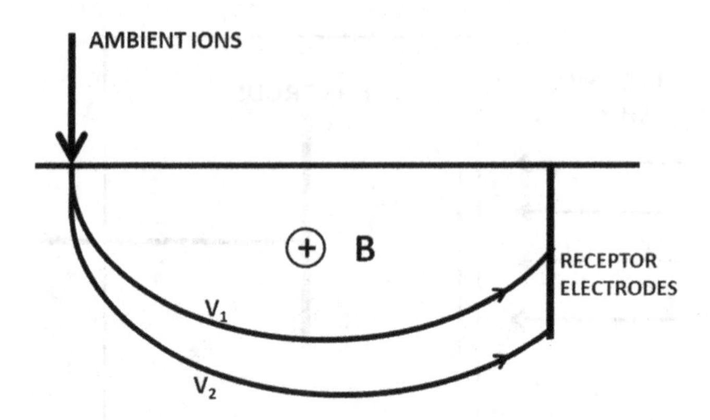

FIGURE 4.5 Plasma analyzer. The idea is the same as a mass spectrometer, but the charged particles have the same mass with different velocities.

Particles of different energies can be received by different receptor electrodes. To save weight, one uses a single receptor only, but the electric, or magnetic, field is varied in discrete steps [Figure 4.5].

4.2.4 Long Booms

Instead of measuring particle energies, experiments have been done to measure the potential difference $\Delta\phi$ between the tip of a long boom and the satellite body. The idea is that the tip of the boom is so far away from the charged body that it is essentially at the space plasma potential ϕ_p, which is zero.

$$\Delta\phi = \phi_S - \phi_p \approx \phi_S \tag{4.10}$$

However, the boom tip itself can charge to a high potential ϕ_p so that the measurement $\Delta\phi$ may not indicate an accurate potential ϕ_S of the satellite body [*Lai*, 1998].

EXERCISES

1. Suppose there is a potential barrier of −2 V at the position r_s = 1 cm outside a satellite at Sun angle $\theta = 0°$. The satellite radius is r_0, and the surface potential at ($r = r_0$, $\theta = 0°$) is +6 V. Figure 4.6 shows the potential profile as a function of distance. At r_0, the potential is −6 V. At $r = r_S$, the potential is +2 V.

   Consider the ambient Maxwellian ions and electrons moving toward the satellite. If you measure the ion and electron energy distributions on the satellite surface at $r = r_0$, $\theta = 0°$, what would the distributions look like?

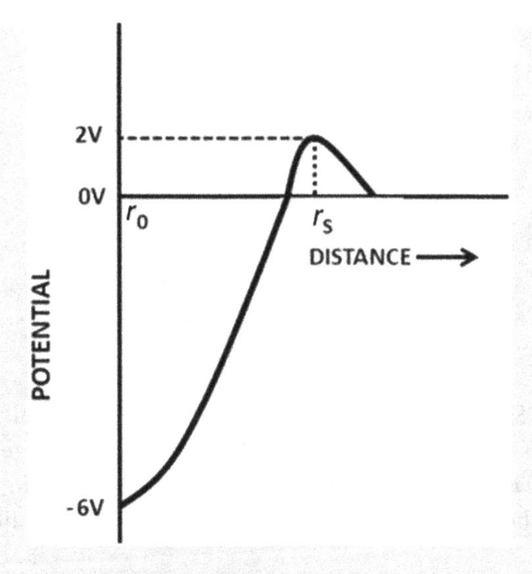

FIGURE 4.6 Potential profile as a function of distance. At r_0, the potential is −6 V; at r_S, the potential is +2 V; at infinity, the potential is 0 V. The ions and electrons are coming in from the right-hand side toward the satellite surface, r_0.

Solution:

The Maxwellian distribution $f(E)$ as a function of energy E is given as follows:

$$f(E) = n\left(\frac{m}{2\pi kT}\right)^{3/2} \exp\left(-\frac{E}{kT}\right) \tag{E4.1}$$

The log function of $f(E)$ is as follows.

$$\log f(E) = \log\left[n\left(\frac{m}{2\pi kT}\right)^{3/2}\right] - \frac{E}{kT} \tag{E4.2}$$

In Figure 4.7, we represent the graph of log $f(E)$ versus energy E. Let us consider incoming ions. When the potential is zero, the distribution is at ABC. At r_S, the potential is +2 V, which is repulsive and therefore slows down the ions, so the distribution shifts to $A_1B_1C_1$. After passing the point r_S, the ions reach the satellite surface, r_0. There, the potential is −6 V, which is attractive, so the distribution becomes $A_2B_2C_2$.

2. Calculate the ratio of A_2C_2 to that at AC.
3. What is the energy difference between A_1 and A_2.
4. Repeat the above Exercises 1–3 for incoming electrons.

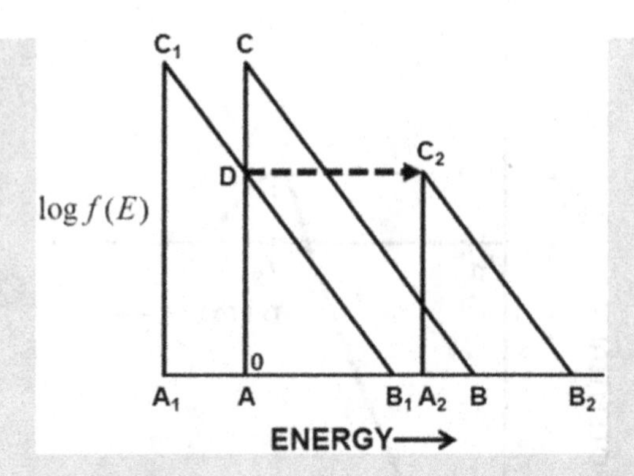

FIGURE 4.7 Shifts of the ion distribution. If the potential is zero, the distribution is at ABC. If the potential is +2 V, which is repulsive and therefore slows down the ions, the distribution shifts to $A_1B_1C_1$. If the potential becomes −6 V, which is attractive, the distribution becomes $A_2B_2C_2$.

5. A long boom extends outside the sheath of a spacecraft charged in a space plasma environment. A common assumption is that the potential ϕ_{boom} of the tip of the boom is at the plasma potential. Is this assumption good? Why or why not?

REFERENCES

Baragiola, R.A., Principles and mechanisms of ion induced electron emission, *Nucl. Instr. Methods in Phys. Res., B.* Beam Interactions with Materials and Atoms, Vol. 78, 223–238, (1993), 10.1016/0168-583X(93)95803-D.

Lai, S.T., A critical overview of measurement techniques of spacecraft charging in space plasma, in *Measurement Techniques in Space Plasmas: Particles*, Geophysical Monograph 102, R.F. Pfaff, J.E. Borovsky, and D.T. Young (eds.), American Geophysical Union, Washington, DC, pp. 217–221, (1998).

Lai, S.T., and C. Miller, Retarding potential analyzer: Principles, designs, and space applications, *AIP Advances*, Vol. 10, 10.095324–10.095328, (2020), 10.1063/5.0014266.

Lin, Y., and D.G. Joy, A new examination of secondary electron yield data, *Surf. Interfaces Anal.*, Vol. 37 (11), 895–900, (2005).

Vampola, A.L., Measuring energetic electrons – What works and what doesn't, in *Measurement Techniques in Space Plasmas: Fields, Geophysical Monograph 103*, R.F. Pfaff, J.E. Borovsky, and D.T. Young (eds.), American Geophysical Union, Washington, DC, pp. 339–355, (1998).

CHAPTER 5

Secondary and Backscattered Electrons

In Chapter 1, we learned that the ambient electron current is much greater than the ambient ion flux because of the mass difference between electrons and ions. Therefore, an object placed in space plasma, or laboratory plasma, would be likely to intercept more electrons than ions. As far as the interplay between these two chess pieces, ambient electrons and ambient ions, is concerned, it is indeed true that an object placed in plasma would likely charge to negative potentials. When the situation involves more than these two chess pieces, the object does not necessarily charge to negative potentials.

5.1 SECONDARY ELECTRONS

Secondary electrons are of central importance in spacecraft charging. When an electron impacts on an atom, it may impart some of its energy E to the atom, resulting in excitation or ionization, depending on the energy available. Likewise, when an electron impacts on a solid, the electron may enter to some depth, thereby imparting its energy E to the charged particles in the solid. As a result, secondary electrons may be generated.

DOI: 10.1201/9780429275043-5

For each incoming electron impacting on a solid surface, the number of secondary electrons coming out from the solid surface is a probability $\delta(E)$, which depends on the energy E and the properties of the solid. The probability $\delta(E)$ may be greater or less than unity. For highly elaborate computations, one needs the angles of incidence of the primary electrons also, but such data are unlikely to be available. For most uses, one need not consider the angle of incidence.

The incoming electrons are also called primary electrons. The probability $\delta(E)$ is also called the secondary electron emission coefficient or the secondary electron yield (SEY). It can be measured statistically in the laboratory as the ratio of the number of outgoing electrons divided by the number of incoming electrons. For general references on secondary electrons, see, for example, *Dekker* [1958]; *Darlington and Cosslett* [1972]; *Sanders and Inouye* [1978]; *Scholtz et al. [1996]*; *Cazaux* [2005]; and *Lin and Joy* [2005].

Figure 5.1 shows a typical graph of $\delta(E)$ as a function of primary electron energy E. For most materials, the graph features two crossings E_1 and E_2 and a maximum E_{max}. Typically, E_1 is about 35 to 45 eV, and E_2 ranges from about 700 to 2000 eV, depending on the material. If δ_{max} is below unity, the E_1 and E_2 crossings are absent. In the energy range, E_1 to E_2, $\delta(E) > 1$, meaning there is a good probability for every incoming electron to generate more than one outgoing (secondary) electron. In this range, the outgoing electron flux exceeds the incoming electron flux,

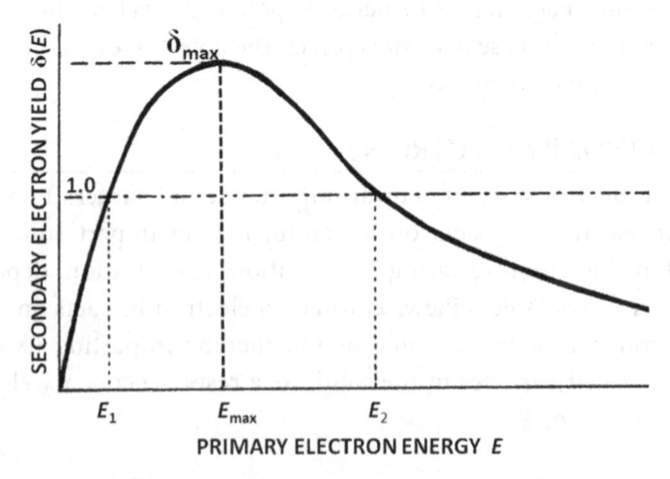

FIGURE 5.1 A typical SEY $\delta(E)$ curve. For most materials, there are two crossings E_1 and E_2 and also a δ_{max} at E_{max}. The $\delta(E)$ curve is monotonically decreasing for E beyond E_{max}.

implying that positive voltage charging is possible. However, the resulting positive voltage is low because secondary electrons have only a few eV in energy typically. If the charging voltage exceeds a few positive volts, the secondary electrons cannot leave.

Physically, if the primary electron is not energetic enough, it probably cannot generate a secondary electron or more. For some materials, δ_{max} can reach 2 or even greater, meaning there can be two or more secondary electrons coming out for every primary electron coming in. If the primary electron is very energetic, it penetrates too deeply into the material so that the secondary electrons generated cannot find their ways out. This is why the graph beyond the maximum is monotonically decreasing.

Since the primary electron shares its energy with the many charged particles, or plasma, modes, onside, the secondary electron has typically a few eV in energy only.

Note also that if the material is very thin, it is possible that the secondary electron may find its way out from the backside of the material.

5.2 BACKSCATTERED ELECTRONS

There is a small probability $\eta(E)$ that the incoming particle may bounce out elastically as a backscattered electron. The probability $\eta(E)$ is also called backscattered electron yield (BEY). Backscattered electrons differ from secondary electrons in two features: (1) the backscattered electron has nearly the same energy as the primary electron, whereas the secondary electrons feature a distribution of energy, and (2) $\eta(E)$ is never greater than unity and is usually small.

Note that in recent years some researchers have pointed out that $\delta(E)$ at very low primary electron energies (a few eV) should be nearly zero because the energy budget needs to overcome the work function. However, experimental data appear to show a nonzero $\delta(E)$ at low energies, and therefore the researchers pointed out that the electrons coming out at low energies are actually backscattered electrons. In other words, $\delta(E)$ may actually be mixed up with $\eta(E)$ in the analysis of experimental data. There is both theoretical and experimental support for this view [*Jablonski and Jiricek*, 1996; *Cazaux*, 2012; *Cimino, et al.*, 2008, 2012].

For spacecraft charging, it is often not necessary to distinguish $\delta(E)$ from $\eta(E)$ – which is very fortunate. For practical purposes, it is sufficient to consider $\delta(E) + \eta(E)$ as the total probability of outgoing electrons from the material surface.

5.3 INCOMING AND OUTGOING ELECTRON FLUXES

The electron flux J arriving at a surface is given as follows.

$$J = qnv \tag{5.1}$$

where q, n, and v are the electron density, electron charge, and electron velocity, respectively. If the electrons are not of the same velocity but form a velocity distribution $f(v)$, the electron flux J is given by an integral over the velocities.

$$J = q\int_0^{\infty} d^3v \; f(v) \; v \tag{5.2}$$

In polar coordinates, the flux J of (5.2) is written as follows:

$$J = q\int_0^{\infty} dv \; v^2 \int_0^{\pi/2} d\varphi \int_0^{2\pi} d\theta \sin\varphi \; f(v) v \tag{5.3}$$

The Maxwellian velocity distribution, $f(v)$, is of the following form:

$$f(v) = n\left(\frac{m}{2\pi kT}\right)^{3/2} \exp\left(-\frac{mv^2}{2kT}\right) \tag{5.4}$$

where m is the electron mass, k the Boltzmann constant, and T the electron temperature. Since electrons are measured as a function of energy E, it is convenient to use E as the variable instead of v. Let us denote $f(E)$ as the electron velocity distribution where $E = (1/2)mv^2$.

$$f(E) = n\left(\frac{m}{2\pi kT}\right)^{3/2} \exp\left(-\frac{E}{kT}\right) \tag{5.5}$$

Using E, the incoming electron flux J in (5.3) is written in the following form:

$$J = q\int_0^{\infty} dE \; E\int_0^{\pi/2} d\varphi \int_0^{2\pi} d\theta \; \sin\varphi \; f(E) \tag{5.6}$$

The balance between the outgoing and incoming electron fluxes can be written as follows:

$$q\int_0^\infty dEE\int_0^{\pi/2} d\varphi\int_0^{2\pi} d\theta\ \sin\varphi\, f(E)$$
$$= q\int_0^\infty dEE\int_0^{\pi/2} d\varphi\int_0^{2\pi} d\theta\ \sin\varphi\, f(E)[\delta(E)+\eta(E)] \qquad (5.7)$$

where $\delta(E)$ and $\eta(E)$ are the secondary and backscattered electron yields. For normal incidence, we need not elaborate the angular dependence of the secondary and backscattered electron yields. Since q and the angles in eq(5.7) cancel out on both sides, the electron flux balance equation becomes [*Lai*, 2013a]:

$$\int_0^\infty dEE\ f(E) = \int_0^\infty dEE\ f(E)[\delta(E)+\eta(E)] \qquad (5.8)$$

This current balance equation, eq(5.8), is very important for spacecraft charging. More on this equation will be discussed in the next chapter.

5.4 EMPIRICAL FORMULAE OF SEY AND BEY

The Sanders and Inouye SEY formula [1978] in eq(5.9) is simple and useful.

$$\delta(E) = c[\exp(-E/a) - \exp(-E/b)] \qquad (5.9)$$

where $a = 4.3E_{max}$, $b = 0.36E_{max}$, and $c = 1.37\delta_{max}$.

The Prokopenko and Laframboise BEY formula [1980] is as follows:

$$\eta(E) = A - B\exp(-CE) \qquad (5.10)$$

where the parameters A, B, and C are material dependent.

There are new SEY and BEY formulae published from time to time. Some of them are obtained from more careful measurements or better data fitting than before. However, one would obtain different results by using different formulae.

Recent work suggests that both SEY and BEY depend not only on the solid material properties but very much on the surface conditions, such as surface smoothness, contamination, cleanliness, scrubbing, temperature, and more.

As Figures 5.2 and 5.3 show, surface conditions can significantly affect the yields, SEY and BEY, from the surfaces. The yields, in turn, can affect the charging of spacecraft (Figure 5.4).

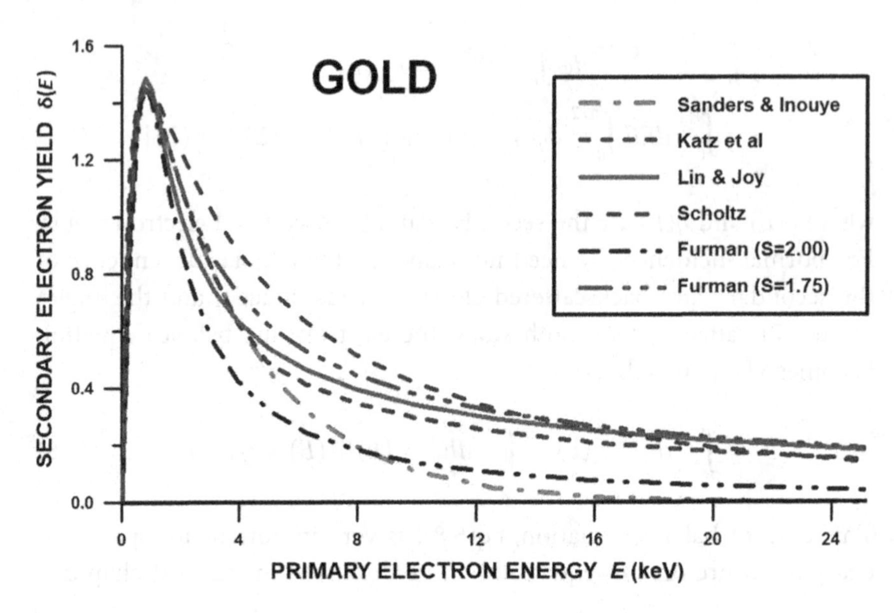

FIGURE 5.2 Results obtained for gold by using different SEY $\delta(E)$ formulae. They are all different. From Lai [2010]. Work of US Government. No copyright.

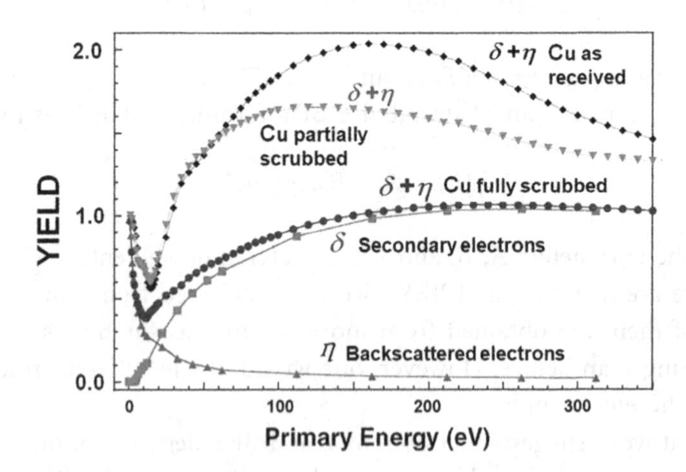

FIGURE 5.3 The *Cimino* [2006] measurements of secondary electron yield, SEY $\delta(E)$, from copper, Cu, at CERN. The backscattered yield $\eta(E)$ is very low except at primary energies below about 40 eV.

Source: From Cimino [2006], Reprinted with permission from Elsevier.

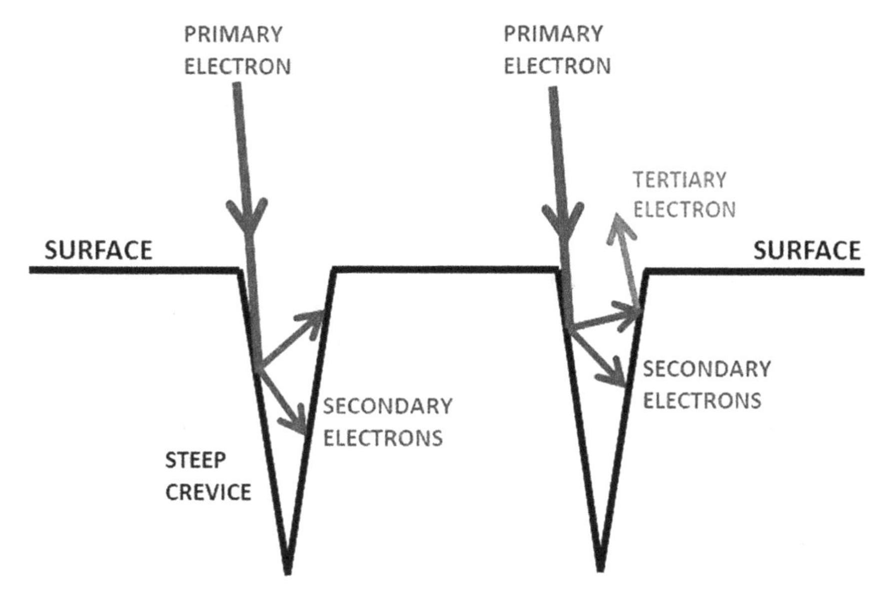

FIGURE 5.4 Reduction of secondary electron yield from a coarse surface. The secondary electrons may impact on the crevice walls. Since secondary electrons have low energies, they have low probability for generating tertiary electrons.

5.5 RESEARCH PROBLEMS

There are research problems of possible interest in this area. SEY is significantly dependent on the surface conditions.

High-energy impact of H^+ ions on surfaces at high (e.g. geosynchronous) altitudes may remove the a priori surface contamination caused by atomic oxygen bombardment at lower altitudes.

EXERCISES

1. Take the Sanders and Inouye SEY formula [1978] provided in eq(5.9). Compute the derivative of the SEY function $\delta(E)$, and then set $d\delta(E)/dE = 0$, which should be satisfied at $E = E_{max}$. Verify that the coefficients a and b are consistent with this derivative-based approach of finding E_{max}.

2. The flux J of incoming electrons or ions can be written as

$$J = \int_0^{\infty} dEE \;\; f(E)$$

where $f(E)$ is the electron or ion distribution. In the above equation, we have neglected the constants π and m, which do not change in spacecraft charging. Suppose we write the above equation as

$$J = \int_0^{\infty} dE \; K(E)$$

where $K(E)$ is called the differential flux.

Suppose one plots a graph of "log $K(E)$ vs E", instead of the usual graph of "log $f(E)$ vs E". Can the properties of ion gap and hidden electron population show up in the "log $K(E)$ vs E" graph?

3. If the secondary electron coefficient (or yield) is enhanced (i.e. δ_{max} increases), does the net electron flux increase or decrease?

REFERENCES

Cazaux, J., A new method of dependence of secondary electron emission yield on primary electron emission energy for application to polymers, *J. Phys. D: Appl. Phys.*, Vol. 38, 2433–2441, (2005).

Cazaux, J., Reflectivity of very low energy electrons (<10 eV) from solid surfaces: Physical and instrumental aspects, *J. Applied Phys.*, Vol. 111, 064903, (2012).

Cimino, R., Surface related properties as an essential ingredient to e-cloud simulations, *Nuclear Instruments and Methods in Phys Res A*, Vol. 561, 272–275, (2006).

Cimino, R., M. Commisso, T. Demma, A. Grilli, M. Pietropaoli, L. Ping, V. Sciarra, V. Baglin, P. Barone, and A. Bonnanno, Electron energy dependence of scrubbing efficiency to mitigate e-cloud formation in accelerarors, Proceedings of EPAC08, Genoa, Italy, (2008).

Cimino, R., M. Commisso, D.R. Grosso, T. Demma, V. Baglin, R. Flammini, and R. Laciprete, Nature of the decrease of the secondary-electron yield by electron bombardment and its energy dependence, *Phys. Rev. Lett.*, Vol. 109, 10.1103, (2012), 10.9.064801.

Darlington, E.H., and V. E. Cosslett, Backscattering of 0.5–10 keV electrons from solid targets, *J. Phys. D: Applied Phys.*, Vol. 5, 1969–1981, (1972).

Dekker, A. J., Secondary electron emission, in *Solid State Physics*, vol. 6, F. Seitz and D. Turnbull, (eds.), Academic Press, New York, (1958).

Jablonski, A. and P. Jiricek, Elastic electron backscattering from surfaces at low energies *Surf. Interface Anal.*, Vol. 24, 781–785, (1996).

Lai, S.T., Spacecraft charging: Incoming and outgoing electrons, Report: CERN-2013-002, pp.165–168, (2013a), ISBN: 978-92-9083-386-4, ISSN: 0007–8328. http://cds.cern.ch/record/1529710/files/EuCARD-CON-2013-001.pdf

Lin, Y., and D.G. Joy, A new examination of secondary electron yield data, *Surf. Interface Anal.*, Vol. 37, 895–900, (2005).

Sanders, N.L. and G.T. Inouye, Secondary emission effects on spacecraft charging: Energy distribution consideration, in Spacecraft Charging Technology 1978, R.C. Finke and C.P. Pike, (eds.), NASA-2071, ADA 084626, pp. 747–755, NASA, Washington, D.C., (1978).

Scholtz, J.J., Dijkkamp, D., and Schmitz, R.W.A., *Philips J. Res.*, Vol. 50 (3-4), 375–389, (1996).

CHAPTER 6

Critical Temperature for the Onset of Spacecraft Charging

6.1 CURRENT BALANCE AT THE ONSET OF SPACECRAFT CHARGING

In Chapter 1, we pointed out that in plasmas the electron flux well exceeds the ion flux. We will use this property in this chapter. Let us consider a scenario in which an uncharged spacecraft is in a space environment of low electron and ion energies. The ambient electrons and ions are not mono-energetic, and they form distributions. Suppose the electron temperature increases gradually. Eventually, spacecraft charging onset occurs. A physical and mathematical explanation of this process will be discussed as follows.

Before the onset of charging, the ambient electron flux greatly exceeds the ambient ion flux, as discussed in Chapter 1. Later, after the onset of negative voltage charging, the positive ions, attracted by the negative spacecraft potential, will provide increasing ion flux as the magnitude of charging increases. But now, before the onset of charging, we can ignore the ion flux.

We now consider the current balance between the current of the incoming ambient electrons and that of the outgoing secondary and backscattered electrons.

DOI: 10.1201/9780429275043-6

$$\int_0^\infty dE\ E\ f(E) = \int_0^\infty dE\ E\ f(E)[\delta(E) + \eta(E)] \tag{6.1}$$

which is the same as eq (5.8). For simplicity, we have assumed that the currents are normal to the surface. The left-hand side is the incoming electron flux and the right-hand side the outgoing electron flux. We use the SEY $\delta(E)$ formula of *Sanders and Inouye* [1978]:

$$\delta(E) = c[\exp(-E/a) - \exp(-E/b)] \tag{6.2}$$

where $a = 4.3E_{max}$, $b = 0.36E_{max}$, and $c = 1.37\delta_{max}$

We also use the BEY $\eta(E)$ formula of *Prokopenko and Laframboise* [1980]:

$$\eta(E) = A - B\exp(-CE) \tag{6.3}$$

where the parameters A, B, and C are material dependent.

It is often a good approximation to use the Maxwellian distribution $f(E)$ for the ambient electrons.

$$f(E) = n\left(\frac{m}{2\pi kT}\right)^{3/2} \exp\left(-\frac{E}{kT}\right) \tag{6.4}$$

where m is the electron mass, k Boltzmann's constant, n the electron density, T the electron temperature, and E the electron energy. These three formulae, eqs (6.2–6.4), are for substitution into eq (6.1), which will have to be integrated over the entire energy E range from 0 to infinity.

6.2 TWO IMPORTANT PROPERTIES

Even before we do the integration over E, we can see two important results already. They are (1) and (2) as follows.

1. Mathematically, since the density n appears on both sides of the equation, n cancels out. This result means that the onset of spacecraft charging is independent of the ambient electron density n. However, this result is counter-intuitive, because, at first thought, it seems obvious that higher electron density would imply more electrons intercepted by the spacecraft.

Physically, this result means that the more electrons are coming in, the more electrons are going out, the proportionally ratio being linear.

Therefore, the current balance between the incoming and outgoing electrons does not depend on the ambient electron density.

2. Mathematically, with the density n cancelled out, what variables are left in eq(6.1)? The energy E is an integration variable but the integral is a definite one, meaning that the energy E is no longer a variable after the integration with definite limits. The other symbols, such as a, b, c, A, B, and C, are constants for a given material surface, while m, k, and π are always constants. Therefore, the equation is a function of temperature T only. Let us denote T^* as the solution that satisfies the current balance equation, eq (6.1).

Physically, if we increase the temperature of the ambient electrons, the slope in Figure 6.1 (Upper) decreases. As the slope decreases,

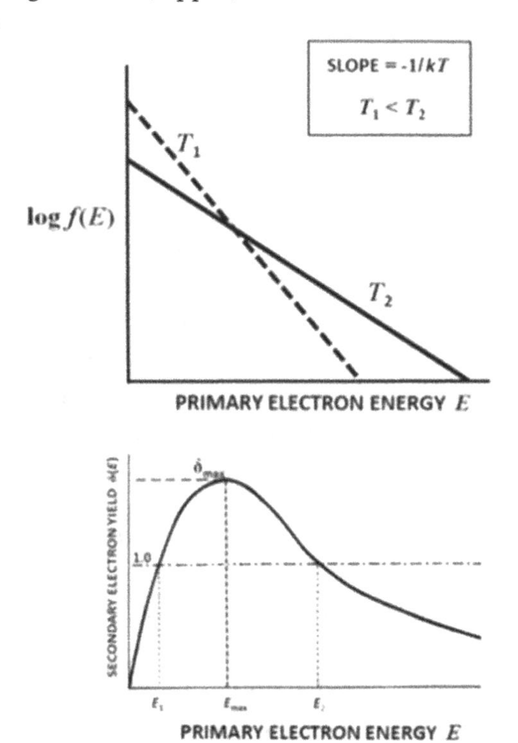

FIGURE 6.1 (Upper) Log of Maxwellian distribution as a function of primary electron energy. The slope is $-1/kT$, where T is the electron temperature. As the temperature T_1 increases to T_2, the slope decreases, resulting in more hot electrons. (Lower) A typical curve $\delta(E)$ of secondary electron yield. It features a maximum δ_{max}, beyond which the curve $\delta(E)$, with $E > E_{max}$, is monotonically deceasing.

there are more and more hot electrons. Since the ambient electrons act as primary electrons as they hit the surface, they generate outgoing secondary electrons. As the temperature T increases, there are more and more hot electrons and fewer and fewer secondary electrons. This is because the SEY $\delta(E)$ decreases monotonically as the primary electron energy E increases. If we keep on increasing the temperature T, there must exist a critical temperature T^* above which there are fewer secondary electrons going out per second than primary electrons coming in.

Another way of illustration is to consider the competition between two camps of electrons. In Figure 6.2 (Upper), the electrons in Camp1 are mostly outgoing, whereas those from Camp2 are mostly incoming. As the temperature T increases, there are more and more hot electrons. The secondary electron yield $\delta(E)$ decreases monotonically as the primary electron energy E increases [Figure 5.2 or 6.2 Lower]. Eventually, at a critical temperature T^*, Camp2 and Camp1 tie in the game. Therefore, beyond T^*, Camp2 wins in overtime, so to speak.

6.3 INTEGRATION RESULT

The current balance model, eq (6.1), in a Maxwellian space environment with the secondary electron yields given by *Sanders and Inouye* [1978] and backscattered electron yields by *Prokopenko and Laframboise* [1980] can produce explicit results of critical temperature T^*. To do so, one substitutes eqs (6.2–6.4) into eq (6.1) and performs the integration. After the integration and a little algebra, the current balance equation, eq (6.1), becomes the following algebraic equation:

$$c\left[(1 + kT/a)^{-2} - (1 + kT/b)^{-2}\right] + A - B(CkT + 1)^{-2} = 1 \quad (6.5)$$

Eq (6.5) can be solved numerically. It can also be solved graphically by plotting the left-hand side as a function of temperature T. The unity crossing of the function gives the critical temperature T^*.

If one prefers to use other secondary electron yield or backscattered electron yield formulae, one can replace eqs (6.2 and 6.3) with the preferred ones. One can also use other distribution functions in place of the Maxwellian distribution. If one wishes to include angle-dependent secondary electron and backscattered electron yields, one needs to include the angle-dependent yields. If one has an anisotropic

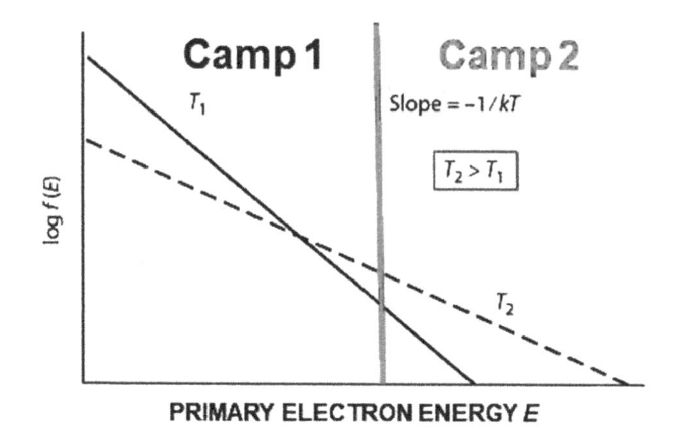

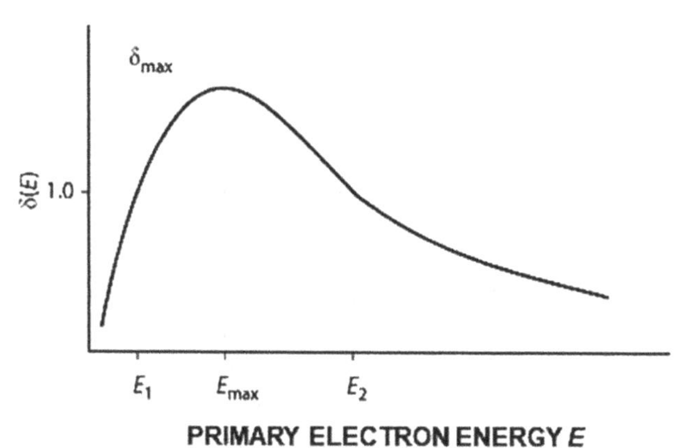

FIGURE 6.2 (Upper) Competition of two camps of electrons. (Lower) A typical SEY $\delta(E)$ curve. In the Upper figure, the electrons in Camp1 are mostly outgoing, whereas those from Camp2 are mostly incoming. As the temperature T increases, there are more and more hot electrons. The secondary electron yield $\delta(E)$ decreases monotonically as the primary electron energy E increases. Eventually, at a critical temperature T^*, Camp2 and Camp1 tie in the game. Therefore, beyond T^*, Camp2 wins in overtime, so to speak.

space plasma distribution function, the integration would be more difficult. We believe that our case, described by eqs (6.1–6.5), is the simplest one for the purpose of bringing out the physics and the results pedagogically.

As an exercise, the reader should derive eq (6.5) from eqs (6.1–6.5). As the second exercise, the reader should practice solving eq (6.5) graphically. In the graph, don't forget to look for the unity crossing, but not the zero crossing.

TABLE 6.1 Critical Temperature for the Onset of Spacecraft Charging

	Critical Temperature T^* (keV)	
Material	**ISO Tropic**	**Normal**
Mg	0.4	–
AL	0.6	–
Kapton	0.8	0.5
Al Oxide	2.0	1.2
Teflon	2.1	1.4
Cu-Be	2.1	1.4
Glass	2.2	1.4
SiO_2	2.6	1.7
Silver	2.7	1.2
Mg Oxide	3.6	2.5
Indium Oxide	3.6	2.0
Gold	4.9	2.9
Cu-Be (Activated)	5.3	3.7
MgF_2	10.9	7.8

6.4 CRITICAL TEMPERATURE FOR VARIOUS MATERIALS

For a given surface material, there exists a critical temperature T^* of the ambient electrons for the onset of spacecraft charging. Using eqs (6.1–6.4), one can calculate the critical temperature T^* for a surface material with known properties of secondary electron yield $\delta(E)$ and backscattered electron yield $\eta(E)$. Table 6.1 shows the critical temperatures for various typical spacecraft surface materials.

Above T^*, the spacecraft potential becomes negative. As the ambient electron temperature further increases, the magnitude of the negative potential of the spacecraft increases. We will discuss how to calculate the magnitude later in this book.

6.5 EVIDENCE OF THE EXISTENCE OF CRITICAL TEMPERATURE

Figure 6.3 shows spacecraft charging data obtained on a Los Alamos National Laboratory (LANL) geosynchronous satellite. The electron temperatures $T_{\|}$ and $T_{\perp}$were measured in directions parallel and perpendicular, respectively, to the ambient magnetic field. The spacecraft potential ϕ is zero when the temperatures are low. One can see an obviously line (dash) of critical temperature T^*, above which the magnitude of the negative potential is finite.

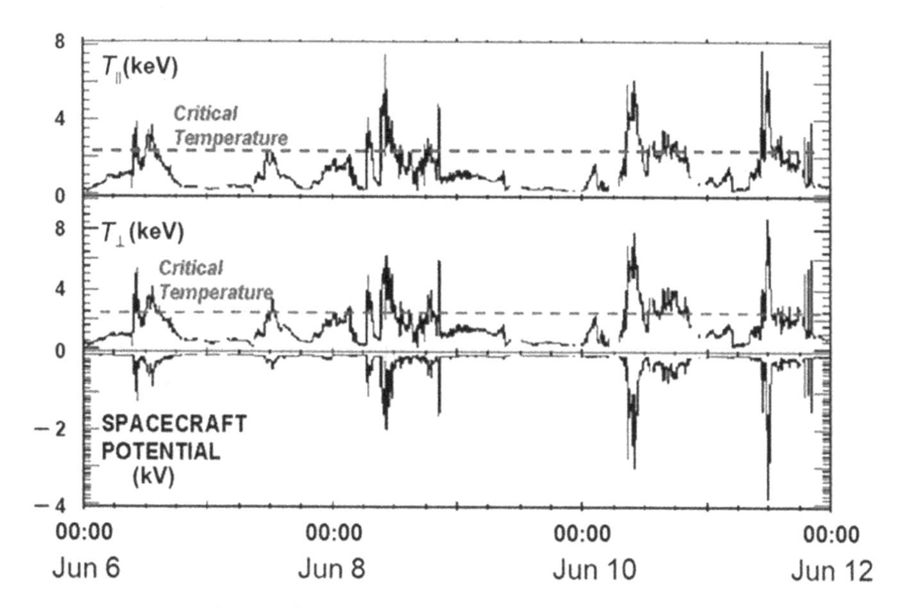

FIGURE 6.3 Evidence of the existence of critical temperature T^*. If the ambient electron temperature T is below T^*, no charging occurs. If T is above T^*, the potential magnitude appears to be approximately proportional to T.

Source: From Lai and Della-Rose [2001]. Work of US Government. No copyright.

Spacecraft charging begins when the ambient electron temperature reaches the critical value T^*. Just above the critical value, or well above the critical value, charging persists. When the ambient electron drops below the critical value T^*, spacecraft charging ceases, which means that the spacecraft potential drops to zero.

As the temperature rises higher above T^*, the potential magnitude goes up higher accordingly. However, as LANL satellite observations show [see, for example, Figure 5a of *Lai*, [2001]], there is no evidence of the charging potential relating to the ambient ion temperature T_i. As discussed in Chapter 1, the ambient electron flux is usually about two orders of magnitude higher than the ambient ion flux. Therefore, as far as the onset of spacecraft charging is concerned, the role of the ambient ions is negligible.

If the negative potential magnitude of the spacecraft becomes higher, the ambient ions will play a more important role in the balance of currents. This important point will be studied in a later chapter.

6.6 BALANCE OF INCOMING AND OUTGOING MAXWELLIAN CURRENTS

Why is it possible to determine the critical temperature T^* by the balance of incoming and outgoing electron currents without considering the ambient ions? It is because the flux of the ambient electrons is about two orders of magnitude higher than that of the ambient ions.

Above the critical temperature T^*, the spacecraft becomes charged to a negative potential, ϕ. Let us examine whether the balance of incoming and outgoing Maxwellian electron currents determine the finite spacecraft potential ϕ without including the role of the ions. In this case, the current balance equation is as follows:

$$\int_0^\infty dE \ E \ f(E + q\phi) = \int_0^\infty dE \ E \ f(E + q\phi)[\delta(E) + \eta(E)] \quad (6.6)$$

For Maxwellian distributions, there is a property:

$$f(E + q\phi) = f(E)\exp(-q\phi) \quad (6.7)$$

Since the exponential terms cancel each other on both sides, the potential ϕ disappears from the current balance equation, eq(6.6). As a result, the potential ϕ is indeterminate mathematically in eq(6.6).

Physically, eq (6.6) describes the incoming electron flux to, and the outgoing electron fluxes from, a charged spacecraft with potential ϕ. The distribution function $f(E)$ is shifted to $f(E+q\phi)$ because of the repulsion by the charging potential. The reason why eq (6.6) cannot have an unique solution of ϕ is because, no matter what the value of ϕ is, the outgoing flux is proportional to the incoming flux. Physically, the more electrons are coming in, the more secondary and backscattered electrons are going out accordingly. Therefore, using the balance of the incoming and outgoing electrons alone cannot determine the spacecraft potential ϕ. We need to include the ions in the current balance for determining the potential.

EXERCISES

1. Using eqs (6.1–6.4), derive the current balance equation in algebraic form:

$$c[(1 + kT/a)^{-2} - (1 + kT/b)^{-2}] + A - B(CkT + 1)^{-2} = 1 \quad (6.5)$$

Solution:

The following shows an explicit mathematical derivation of the current balance equation in algebraic form, which can yield the critical temperature T^* as the solution.

The assumptions taken are (1) the ambient electron distribution $f(E)$ is Maxwellian, (2) the SEY, $\delta(E)$, is given by the *Sanders and Inouye* [1978] formula, and (3) the BEY, $\eta(E)$, is given by the *Prokopenko and Laframboise* [1980] formula. They are eq (6.4), eq (6.2), and eq (6.3), respectively, as follows:

$$f(E) = n\left(\frac{m}{2\pi kT}\right)^{3/2} \exp\left(-\frac{E}{kT}\right) \tag{6.4}$$

$$\delta(E) = c\left[\exp(-E/a) - \exp(-E/b)\right] \tag{6.2}$$

where $a = 4.3E_{max}$, $b = 0.36E_{max}$, and $c = 1.37\delta_{max}$

$$\eta(E) = A - B\exp(-CE) \tag{6.3}$$

where the parameters A, B, and C are material dependent. In carrying out the derivation, let us assume for generality that the ambient electron energy ranges from a lower cutoff E_L to an upper cutoff E_U, which may occur in some geomagnetic situations [*Thomsen, et al.*, 2002]. The complete ambient electron energy range is from zero to infinity.

It is instructive to consider the contributions of the secondary and backscattered electrons separately. The ratio S of the flux of secondary electrons to that of the primary electrons is given by

$$S = \frac{\int_{E_L}^{E_U} dE\, E\ \delta(E)\exp(-E/kT)}{\int_{E_L}^{E_U} dE\ E\ \exp(-E/kT)} \tag{E6.1}$$

Substituting $\delta(E)$ into eq (E6.1), and integrating over E, we have the result:

$$S = c\left[\frac{K(a)}{(1 + k\ T/a)^2} - \frac{K(b)}{(1 + k\ T/b)^2}\right] \tag{E6.2}$$

where

$$K(x) = \frac{\left[(1 + \alpha(x)E)\exp(-\alpha(x)E)\right]_{E_L}^{E_U}}{\left[(1 + E/kT)\exp(-E/kT)\right]_{E_L}^{E_U}} \tag{E6.3}$$

and

$$\alpha(x) = \frac{1}{x} + \frac{1}{kT} \tag{E6.4}$$

For the complete electron energy range (0 to ∞), the ratio S is as follows:

$$\lim_{\substack{E_U \to \infty \\ E_L \to 0}} S = c\,[(1 + kT/a)^{-2} - (1 + kT/b)^{-2}] \tag{E6.5}$$

The ratio R of the flux of backscattered electrons to that of the primary electrons is given by

$$R = \frac{\int_{E_L}^{E_U} dE \;\; E\,\eta(E)\exp(-E/kT)}{\int_{E_L}^{E_U} dE \;\; E\; \exp(-E/kT)} \tag{E6.6}$$

Substituting $\eta(E)$ into eq(E6.6), and integrating over E, we have the result:

$$R = A - \frac{B}{(CkT+1)^2}\frac{[1 + (C + 1/kT)E\exp(-(C + 1/kT)E)]_{E_L}^{E_U}}{[(1 + E/kT)\exp(-E/kT)]_{E_L}^{E_U}} \tag{E6.7}$$

For the complete electron range (0 to ∞), the ratio R is as follows:

$$\lim_{\substack{E_U \to \infty \\ E_L \to 0}} R = A - B/(CkT + 1)^2 \tag{E6.8}$$

We can substitute eq (E6.5) and eq (E6.8) into the current balance equation eq (6.1). For simplicity here, let us assume that the electron energy ranges from 0 to ∞. Thus, adding eqs (E6.5) and (E6.8) together, we obtain the following simple algebraic equation [Lai, 1991, 2001]:

$$c\,[(1 + kT/a)^{-2} - (1 + kT/b)^{-2}] + A - B(CkT + 1)^{-2} = 1 \tag{6.5}$$

Note that eq (6.5) in *Lai and Della-Rose* [2001] had a misprint. The right-hand side should be unity as shown above. However, the misprint in 2001 did not affect the rest of that paper.

2. Using the current balance equation, eq (6.5), find the critical temperatures of the following surface materials:

Material	E_{max}(keV)	δ_{max}	Z	A	B	C
Aluminum	0.30	0.97	13	0.16	0.30	0.34
Gold	0.80	1.45	79	0.48	0.36	0.61
Teflon	0.30	3.00	8	0.09	0.00	0.00
Indium oxide	0.80	1.40	24.4	0.28	0.08	0.54
MgF_2	0.85	6.38	10	0.13	0.02	0.34

REFERENCES

Lai, S.T., Space plasma temperature effects on spacecraft charging, 39th AIAA Aerospace Science Mtg, Reno, NV, AIAA-2001-0404, (2001).

Lai, S.T., Spacecraft charging thresholds in single and double Maxwellian space environments. *IEEE Trans. Nucl. Sci.*, Vol. 19, 1629–1634, (1991).

Lai, S.T., and D. Della-Rose, Spacecraft charging at geosynchronous altitudes: New evidence of the existence of critical temperature. *J. Spacecraft Rockets*, Vol. 38 (6), 922–928, (2001).

Laciprete, R., D.R. Grosso, R. Flammini, and R. Cimino, The chemical origin of SEY at technical surfaces, in eCloud'12 Proceedings, Joint INFN-CERN-EuCARD-AccNet Workshop on Electron-Cloud Effects, ed. By R. Cimino, G. Rumolo and F. Zimmermann, CERN-2013-002 (CERN, Geneva, 2013), pp. 99–104, DOI:10.5170/CERN-2013-002.

Prokopenko, S.M.L., and J.G. Laframboise, High-voltage differential charging of geostationary spacecraft, *J. Geophys. Res.*, Vol. 85 (A8), 4125–4131, (1980).

Sanders, N. L., and G. T. Inouye, Secondary emission effects on spacecraft charging: Energy distribution considerations, in *Spacecraft Charging Technology*, R.C. Finke and C.P. Pike (eds.), NASA-2071, ADA-084626, Air Force Geophysics Laboratory, Hanscom AFB, MA, pp. 747–755, (1978).

Thomsen, M. F., Korth, H., & Elphic, R. C. Upper cutoff energy of the electron plasma sheet as a measure of magnetospheric convection strength. *J. Geophys. Res.*, Vol. 107(A10), 1331, 10.1029/2001JA000148, (2002).

CHAPTER 7

Importance of Surface Conditions

7.1 MAIN REASONS FOR INACCURACY

The critical temperatures of various surface materials listed in Table 6.1 offer good approximation but not highly accurate values. There are several reasons for the insufficiency in accuracy. A main reason for the insufficient accuracy is that the available empirical secondary electron yield $\delta(E)$ formulae for various materials give different results. Another main reason is that the secondary electron yields of various materials depend very much on the surface conditions. We need to know the surface conditions accurately if we want to have accurate yields for calculating the critical temperatures and the spacecraft potentials.

There are possibly other reasons that will emerge as spacecraft charging research advances. The importance of such advances will hopefully be appreciated in the future. At present, let us discuss surface conditions as follows.

7.2 SECONDARY ELECTRON YIELD FORMULAE

Empirical formulae are obtained by fitting the data obtained from experiments, which are usually very carefully conducted. In recent years, it has been found that the conditions of the material surfaces play significant role in the secondary electron yield (SEY).

Some of the best, and commonly used, empirical secondary electron yield $\delta(E)$ formulae are, for example, those of *Sanders and Inouye* [1978],

DOI: 10.1201/9780429275043-7

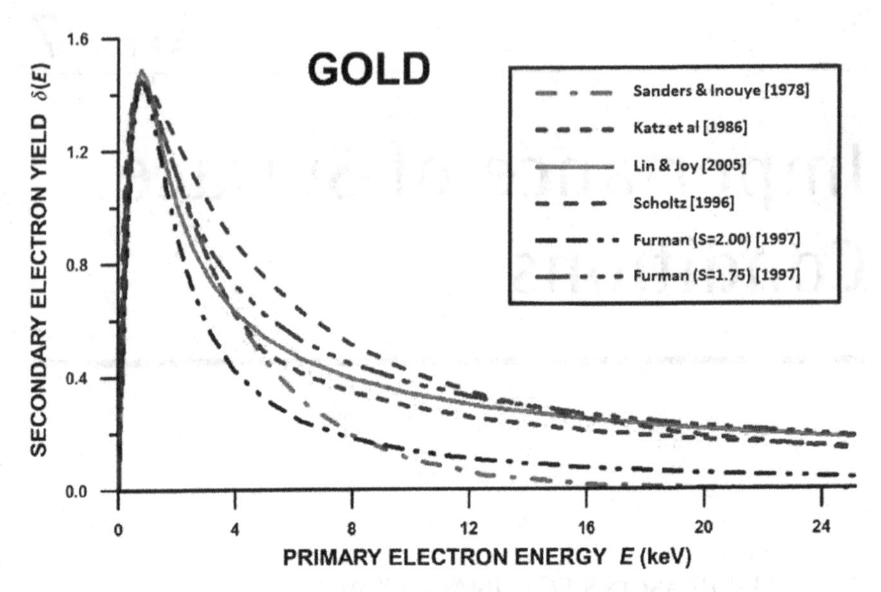

FIGURE 7.1 Secondary electron yield $\delta(E)$ from gold surface obtained by using the empirical formulae of *Sanders and Inouye* [1978], *Katz, et al.* [1986], *Scholtz, et al.* [1996], and *Lin and Joy* [2005]. The results obtained by using *Furman, et al.* [1997] with two different surface parameters are also shown.

Source: From Lai [2010]. Work of US Government. No copyright.

Katz, et al. [1986], *Scholtz, et al.* [1996], and *Lin and Joy* [2005]. If one uses these formulae to calculate the SEY of a given material, they give different results, although, remarkably, all these formulae give the location E_{max} and the amplitude δ_{max} of the SEY $\delta(E)$ graph correctly. Figure 7.1 shows the results of $\delta(E)$ obtained from the above four references for the surface material gold. The results of using *Furman, et al.* [1997] formula with two different surface parameters are also shown. It is remarkable that the results using different $\delta(E)$ formulae are all different.

As a result, the critical temperature T^* values calculated by using different formulae of SEY $\delta(E)$ are all different, as shown in Figure 7.2. It is a good question to ask: Which one of the SEY $\delta(E)$ formulae available should one choose as the best one? One needs to know the surface conditions.

7.3 SURFACE CONDITIONS

Chemical contamination of surfaces can greatly affect the SEY [Laciprete, et al., 2013]. Experiments conducted at CERN, Geneve, Switzerland, found that surfaces after scrubbing have significant difference in SEY

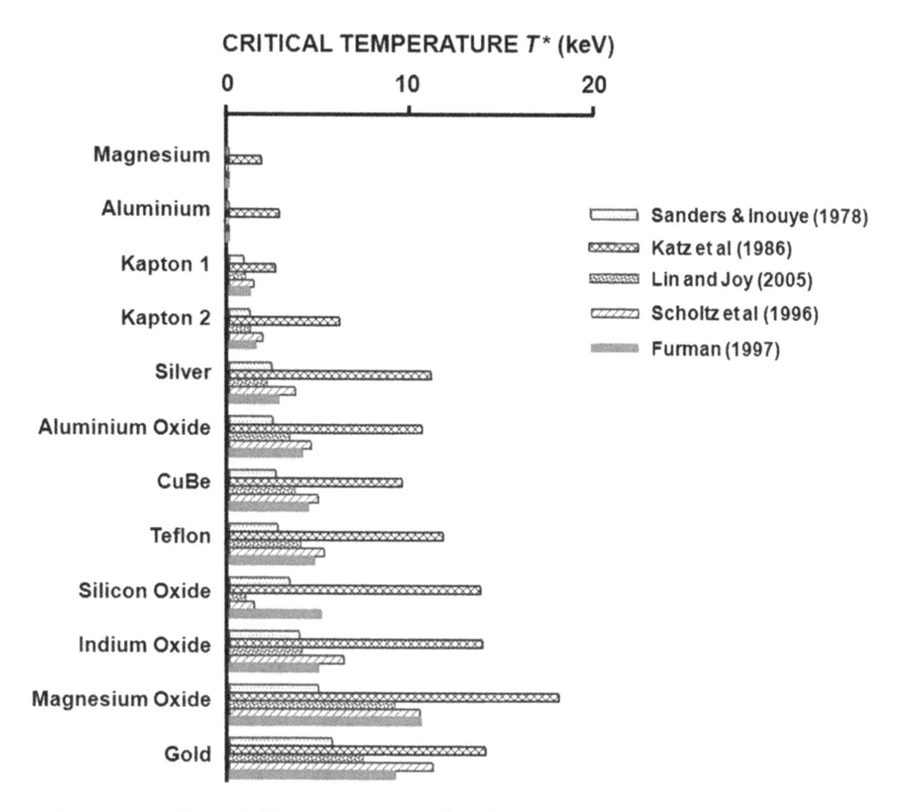

FIGURE 7.2 Critical Temperature T^* for the onset of spacecraft charging calculated by using various best $\delta(E)$ functions. The surface parameter used here in the Furman formula is $S = 1.75$. From Lai [2010]. Work of US Government. No copyright.

compared with surfaces before scrubbing [*Cimino, et al.*, 2012]. CERN is very careful about SEY because their acceleration tubes, which are many miles long, may have secondary electrons generated inside by electron impact or by bremsstrahlung. The secondary electrons may disturb the trajectories of the charged particles being accelerated [*Cimino, et al.*, 2004].

If a surface is not uniformly flat, the SEY $\delta(E)$ is different from that of a uniformly flat surface. For example, if the surface is smoothly wavy, the effective surface area is larger than that of a uniformly flat surface [Figure 7.3]. The chance of receiving a primary electron is slightly enhanced, and a secondary electron may find its way out and escape easier. As a result, the SEY $\delta(E)$ may be enhanced.

On the contrary, a surface with many scratches may reduce the SEY $\delta(E)$. This is obvious from Figure 7.4 as an example. The secondary

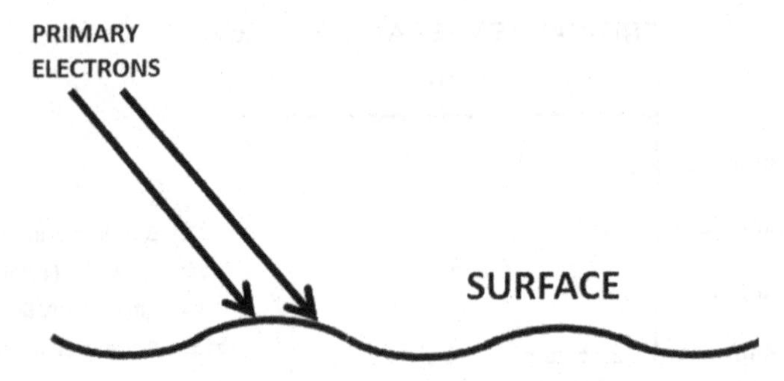

FIGURE 7.3 A gentle wavy surface. The effective surface area is larger than a flat surface, resulting in an increase in secondary electron yield $\delta(E)$.

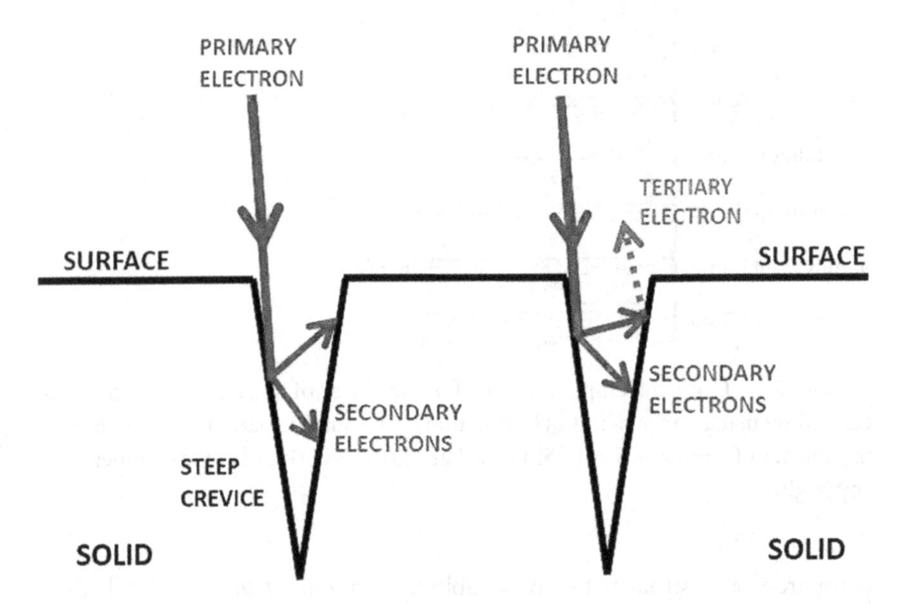

FIGURE 7.4 A surface with crevices. Some secondary electrons are absorbed. They may not be energetic enough to generate tertiary electrons. This surface condition reduces the secondary electron yield $\delta(E)$.

electron generated in a scratch, or crevice, hits the wall of the crevice. Can the secondary electron generate a tertiary electron? Not likely! This is because in doing so, the secondary electron acts as a primary electron for generating the tertiary electron. The secondary electron probably has energy of a few eV to about 15 eV. The probability $\delta(E)$ for generating a secondary electron for a primary electron with an energy in the range of a few eV to 15 eV is very low [see Figure 5.1]. Therefore, the secondary electron generated inside the crevice is practically absorbed without emitting another electron of the next generation (tertiary). As a result, the SEY $\delta(E)$ of scratchy surfaces is reduced.

The Large Hadron Collider (LHC) of CERN has adopted the Furman formula (1997) for SEY $\delta(E)$. The formula features an empirical surface condition parameter s, which one can obtain from the actual measurements of the SEY $\delta(E)$ of the surface material to be used. The Furman formula (1997) is as follows:

$$\delta(E) = \delta_{max} \frac{s(E/E_{max})}{s - 1 + (E/E_{max})^s} \tag{7.1}$$

$$E_{max}(\theta) = E_{max}(0)[1 + 0.7(1 - \cos\theta)] \tag{7.2}$$

$$\delta_{max}(\theta) = \delta_{max}(0)\exp[0.5(1 - \cos\theta)] \tag{7.3}$$

where $90° \geq \theta \geq 0°$ is the primary electron incidence angle measured relative to the normal to the surface. The empirical parameter s characterizes the condition of a given surface material and significantly affect the SEY $\delta(E)$ graph. As an example, the SEY $\delta(E)$ graphs for $s = 1.3$, 1.45, 1.60, 1.75, and 1.90, are shown in Figure 7.5 using Furman's formula, eq (7.1).

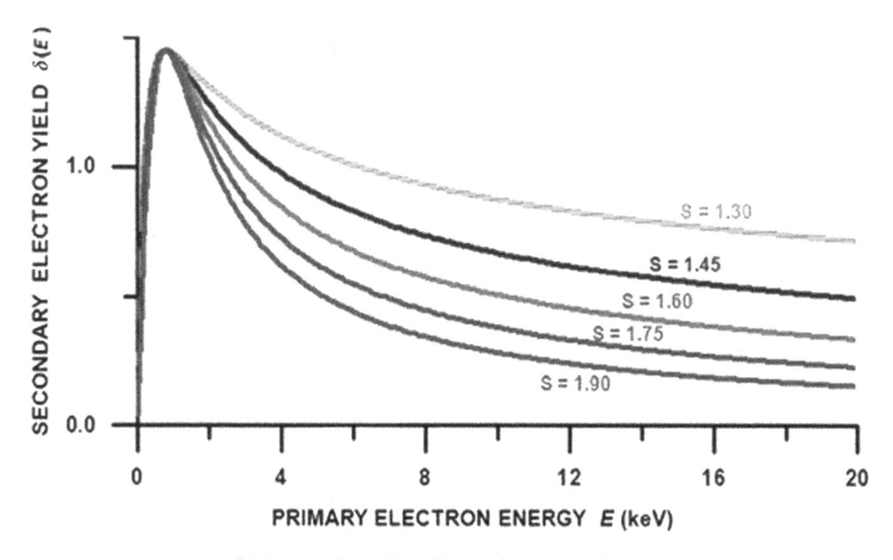

FIGURE 7.5 SEY $\delta(E)$ graphs of gold surface calculated by using Furman's formula (1997). The graphs with different surface parameter s are all different, while the maximum δ_{max} and the location E_{max} of the maximum are unaffected. From Lai [2010]. Work of US Government. No copyright.

7.4 BACKSCATTERED ELECTRON YIELD

Backscattered electron yield (BEY), η, is usually much smaller than SEY. Typically, η is about 0.1 or less. It is often a good approximation to neglect it. However, in recent years, it was found that η, for primary electron energy between 0 to about 25 eV, rises toward unity as the primary electron energy decreases toward zero [see Figure 5.3]. There are experimental [*Cimino, et al.* 2004; *Cimino*, 2006] and quantum calculations [*Jablonski and Jiricek*, 1996] to support this behavior of BEY at low energies.

For spacecraft charging applications, the modification of η at low energies (0–25 eV) may not be necessary for geosynchronous satellites because the electron energy there is usually high (keV). For low energy spacecraft applications, such as the anti-critical temperature [*Lai and Tautz*, 2008; Garnier, et al. 2013], the modification in η is needed.

7.5 APPLICATIONS TO SPACECRAFT SURFACE CONDITIONS

For spacecraft charging, perhaps a piece of test material with surface conditions identical to that used on the spacecraft can be used for measurement of SEY $\delta(E)$ under a simulated space condition in the laboratory. Even so, there are uncertainties in the surface conditions of the actual spacecraft material because the space environment may change the conditions, for example, atomic oxygen bombardment in the ionosphere, high energy electrons and ions bombarding, or even penetrating, the spacecraft surface, especially during extreme geospace events.

Atomic oxygen may change the chemical composition of the surface top layer. Atomic hydrogen bombardment at geosynchronous altitudes may remove the oxidized layer, but this has to be demonstrated in the laboratory under simulated condition together with theoretical support, which seems to be lacking at the time of writing.

High-energy ion impact in space may also cause coarseness on surfaces, especially as time goes by. As a result, microscopic crevices may form on them, altering the SEY $\delta(E)$.

Also, a piece of satellite surface may experience alternating "'bright sunlight and dark shadow" thermal stress as the satellite rotates. As a result, the surface may depart from its original geometry to become a wavy one, and, again, aging is a factor.

Also, if discharges occur from time to time between surfaces of different materials, the conditions of such surfaces may change, altering the SEY $\delta(E)$.

The above problems may offer an opening for research opportunities.

EXERCISES

1. Is reduction of SEY good for lowering the level of spacecraft charging?
2. Suggest a nonspace application where reduction of SEY is necessary.

REFERENCES

Cimino, R., Surface related properties as an essential ingredient to e-cloud simulations, *Nucl. Instr. Methods Phys. Res. A*, (561), 272–275, (2006), 10.1016/j.nima.2006.01.042

Cimino, R., I.R. Collins, M.A. Furman, M. Pivi, F. Ruggerio, G.Rumulo, and F. Zimmermann, Can low-energy electrons affect high-energy physics accelerators?, *Phys. Rev. Lett.*, Vol. 93(1), 14801–14804, (2004).

Cimino, R., M. Commisso, D.R. Grosso, T. Demma, V. Baglin, R. Flammini, and R. Laciprete, Nature of the decrease of the secondary-electron yield by electron bombardment and its energy dependence, *Phys. Rev. Lett.*, Vol. 109, (2012), 10.1103/PhysRevLett.109.064801

Furman, M. et al., MBI 97 Tsukuba, KEK 97–17, (1997).

Garnier, P., M.K.G. Holmberg, J.-E. Wahlund, G.R. Lewis, S.R. Grimald, M.F. Thomsen, D.A. Gurnett, A.J. Coates, F.J. Crary, and I. Dandouras, (2013), The influence of the secondary electrons induced by energetic electrons impacting the Cassini Langmuir probe at Saturn, *J. Geophys. Res. Space Physics*, 118, 7054–7073, doi:10.1002/2013JA019114.

Jablonski, A., and P. Jiricek, Elastic electron backscattering from surfaces at low energies, *Surf. Interface Anal.*, Vol. 24, 781–785, (1996).

Katz, I., M. Mandell, G. Jongeward and M.S. Gussenhoven, The importance of accurate secondary electron yields in modeling spacecraft charging, *J. Geophys. Res.*, Vol. 91 (A12), 13739–13744, (1986).

Lai, S.T., The importance of surface conditions for spacecraft charging, *J. Spacecraft & Rockets*, Vol. 47 (4), 634–638, (2010), 10.2524/1.48824.

Lai, S.T., and M. Tautz, On the anti-critical temperature in spacecraft charging, *J. Geophys. Res.*, Vol. 113, A11211, (2008), 10.1029/2008JA01361.

Lin, Y., and D.G. Joy, A new examination of secondary electron yield data, *Surf. Interfact Anal.*, Vol. 37, 895–900, (2005).

Sanders, N.L., and G.T. Inouye, Secondary emission effects on spacecraft charging: Energy distribution consideration, in Spacecraft Charging Technology 1978, R.C. Finke and C.P. Pike (eds.), NASA-2071, ADA 084626, pp. 747–755, NASA, Washington, D.C., (1978).

Scholtz, J.J., D. Dijkkamp, and R.W.A. Schmitz, Secondary electron properties, *Philips J. Res.*, Vol. 50 (3–4), 375–389, (1996).

CHAPTER 8

High-Level Spacecraft Potential

8.1 BEYOND THE CRITICAL TEMPERATURE

If the ambient electron temperature rises beyond the critical temperature, there are, of course, more and more ambient hot electrons. The flux of hot electrons increases accordingly. As a result, the magnitude of the negative spacecraft potential ϕ rises. The ambient positive ions are therefore attracted to the spacecraft. In this case, the ion current is not negligible; it becomes an additional chess piece in the current balance:

$$I_T(\phi) = I_e(0)[1 - <\delta + \eta>]\exp\left(-\frac{q_e\phi}{k\,T_e}\right) - I_i(0)\left(1 - \frac{q_i\phi}{k\,T_i}\right)^{\alpha} = 0 \quad (8.1)$$

where

$$<\delta + \eta> = \frac{\int_0^\infty dE\ \ E\ f(E)[\delta(E) + \eta(E)]}{\int_0^\infty dE\ \ E\ f(E)} \neq 1 \quad (8.2)$$

In eq (8.1), $I_T(\phi)$ is the total current, $I_e(\phi)$ the ambient electron current, $\phi(<0)$ the spacecraft potential, $q_e(<0)$ the charge of an electron, $q_i(>0)$ that of a positive ion, k the Boltzmann constant, T_e the electron temperature, and T_i the ion temperature. The last term in parenthesis with

DOI: 10.1201/9780429275043-8

an α-exponent is positive because $\phi < 0$. The last term is known as the *Mott-Smith and Langmuir* [1926] orbit-limited attraction term, where α = ½ for a long cylinder, 1 for a sphere, and 0 for a large plane, approximately. Using the electron beam method, the exponent α for the SCATHA satellite was found to be α = 0.75 [*Lai*, 1994].

Eq (8.2) is the normalized outgoing current of secondary and backscattered electrons. The denominator is the incoming ambient current $I_e(0)$.

$$I_e(0) = \int_0^{\infty} dE \ E \ f(E) \tag{8.3}$$

Current balance requires that the total current $I_T(\phi) = 0$. If one solves eq (8.1), the solution is the spacecraft potential ϕ, which is negative. One way to solve the equation is to plot the graph of the total current $I_T(\phi)$, which is the left-hand side of eq (8.1), as a function of the variable ϕ on the x-axis in Figure 8.1. The graph crosses the x-axis at a value of ϕ, which is the spacecraft potential.

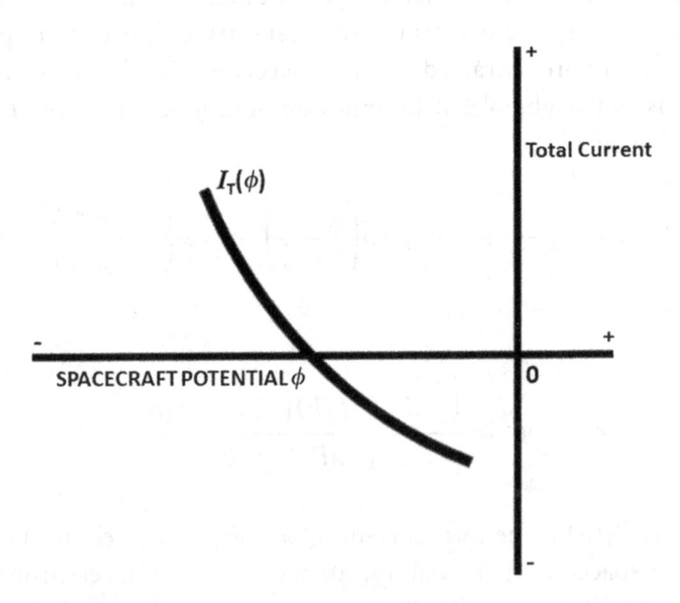

FIGURE 8.1 Graphical solution of the current balance equation $I_T(\phi) = 0$. If the spacecraft potential ϕ is negative and there are more positive ions coming to the spacecraft, $I_T(\phi)$ is positive and on the upper half of the *x*-*y* plane. If there are more electrons than ions coming in, $I_T(\phi)$ is on the lower half plane. The spacecraft potential ϕ is at $I_T(\phi) = 0$, where the graph crosses the x-axis.

8.2 ION-INDUCED SECONDARY ELECTRONS

Below the critical temperature, the spacecraft is uncharged, and therefore the ambient electron flux exceeds that of the ambient ions by nearly two orders of magnitude [see Figure 1.4]. Above the critical high temperature, the spacecraft is charged by the high-energy ambient electrons to negative potentials. See Figure 8.2 for the detail. Highly negative potentials ϕ attract the ambient positive ions, as described by the ion term with an exponent α in eq (8.1). The ion term is positive and greater than unity, because the ion charge q_i is positive and the potential ϕ is negative. As a result, the ion flux increases as the spacecraft potential ϕ increases. Note that there is an ion temperature T_i in the denominator of the last term in eq (8.1), meaning that as the ion temperature T_i increases, the ion flux decreases. This is because of the *Mott-Smith and Langmuir's* [1926] orbit-limited theorem [see Figure 3.2], which implies that the attracted ions may actually miss the central target if their angular momenta are too high, for these ions are too fast or too far away.

The secondary electron yield ISEY $\gamma_i(E)$ due to positive ion impact is given by [*Katz, et al.*, 1977; *Whipple*, 1981]:

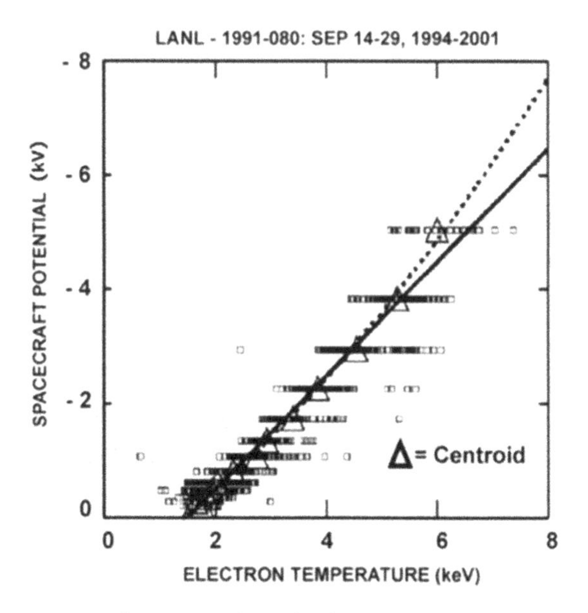

FIGURE 8.2 Spacecraft potential and electron temperature measured on Spacecraft LANL-1990-095, during eclipse periods, September 14–29, 1993–2001. A centroid is the average temperature of all the points on the same voltage level.

Source: From Sec. 5 in Lai and Tautz [2006].

$$\gamma_i(E) = 2\gamma_m \frac{(E/E_m)^{1/2}}{(1 + E/E_m)} \tag{8.4}$$

where γ_m is the maximum ISEY yield at an energy E_m. For example, the ISEY $\gamma_i(E)$ for aluminum oxide Al_2O_3 is, according to eq (8.4), given by

$$\gamma_i(E) = 1.36 \frac{E^{1/2}}{1 + E/40} \tag{8.5}$$

Including the ion-induced secondary electrons, the total current eq (8.1) becomes

$$I_T(\phi) = I_e(0)[1 - <\delta + \eta>]\exp\left(-\frac{q_e\phi}{kT_e}\right) - I_i(0)\left(1 - \frac{q_i\phi}{kT_i}\right)^{\alpha}[1 + <\gamma_i>] \tag{8.6}$$

where the normalized electron-induced secondary electron current $<\delta + \eta>$ is given by eq (8.2), and the normalized ion-induced secondary electron current is given as follows:

$$< \gamma_i > = \frac{\int_0^\infty dE\ E\ f_i(E)\gamma_i(E)}{\int_0^\infty dE\ E\ f_i(E)} \tag{8.7}$$

where $f_i(E)$ is the ion distribution function, $I_i(\phi)$ the ambient ion current, and ϕ the spacecraft potential.

$$I_i(0) = \int_0^\infty dE\ E\ f_i(E) \tag{8.8}$$

For some recent references on ion-induced secondary electron yield, see, for example, *Hasselkamp, et al.* [1986], *Rajopadhye, et al.* [1986], and *Baragiola* [1993].

8.3 KAPPA DISTRIBUTION

At energies well beyond the critical temperature, the space plasma often deviates from being Maxwellian. Let us examine some data from the LANL satellites. In Figures 3–5 of *Lai and Tautz* [2006], the spacecraft potential deviates upward from a straight line starting approximately at electron energies above 5 keV. In comparison, eq (8.1) shows that as the negative potential increases to higher values, the electrons are increasingly repelled and the ion current becomes more important. If the spacecraft resembles somewhere between sphere to a cylinder, then the exponent α is about 1 to ½. Yet, the observed data in *Lai and Tautz* [2006] deviates upward. The upward deviation cannot be explained by ion-induced secondary electrons because they would bend the curves in Figures 3–5 of *Lai and Tautz* [2006] downward. Therefore, the observed data in the figures hint at a possibility that the distribution is no longer Maxwellian at electron energies higher than about 5 keV.

It appears possible that the distribution becomes one resembling a Maxwellian plus a high-energy tail. In this manner, it is often a good approximation to describe the distribution as a Kappa distribution $f_\kappa(E)$ [*Vasyliunas*, 1968; *Meyer-Vernet*, 1999], which is as follows:

$$f_\kappa(E) = n\frac{\Gamma(\kappa+1)}{\Gamma(3/2)\Gamma(\kappa-1/2)}\left[\left(\kappa-\frac{3}{2}\right)T_\kappa\right]^{-\frac{3}{2}}\left[1+\frac{E}{(\kappa-3/2)T_\kappa}\right]^{-(\kappa+1)} \tag{8.9}$$

This distribution function is characterized by the kappa temperature T_κ, related to the physical temperature T via the following relation (see Figure 8.3 for graphical representation)

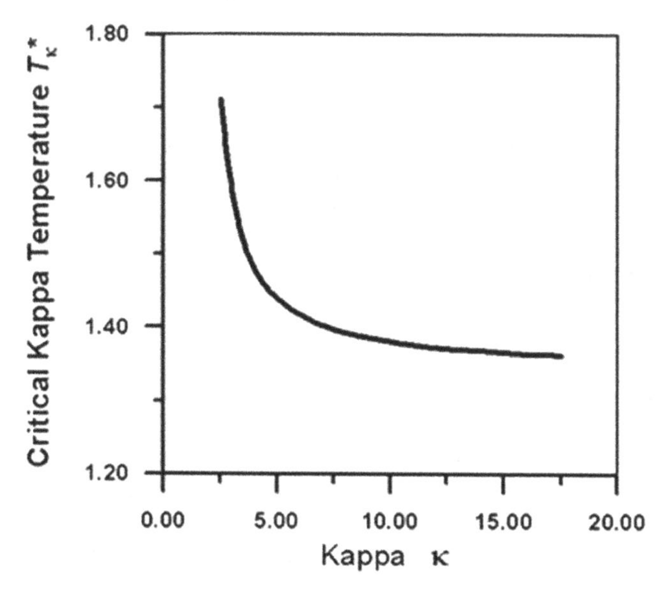

FIGURE 8.3 Critical temperature T_κ^* for copper-beryllium (un-activated) as a function of kappa [*Lai*, 2008].

$$T_\kappa = \frac{\kappa}{(\kappa - 3/2)} T \tag{8.10}$$

where possible values of the kappa parameter are in the range

$$3/2 < \kappa < \infty \tag{8.11}$$

In the limit $\kappa \to \infty$, one recovers the Maxwellian distribution. In order to find the onset of spacecraft charging, the current balance equation is solved by using $f_\kappa(E)$.

$$\int_0^\infty dE\, E\, f_{K,e}(E) \exp\left(-\frac{q_e \phi}{k\ T_e}\right) = \left(\frac{m}{M}\right)^2 \int_o^\infty dE\, E\, f_{K,i}(E)\, [\delta(E) + \eta(E)] \left(1 - \frac{q_i\ \phi}{k\ T_i}\right) \tag{8.12}$$

yielding the critical kappa temperature T_κ^* [*Harris*, 2003; *Lai*, 2008] for each given surface material characterized by the secondary electron yield $\delta(E)$ and backscattered electron yield $\eta(E)$. At the asymptote of $\kappa \to \infty$, one recovers the Maxwellian result T^*. Note also that the density n on both sides of the current balance equation cancels each other out.

EXERCISE

1. If you try to determine the spacecraft potential in a Maxwellian space plasma by balancing the incoming and outgoing electron fluxes, is the potential calculated (a) positive, (b) negative, or (c) indefinite? What ingredient has been missing in this model of current balance?

REFERENCES

Baragiola, R.A., Principles and mechanisms of ion induced electron emission, *Nucl. Instr. Methods in Phys. Res., B.* Beam Interactions with Materials and Atoms, Vol. 78, 223–238, (1993), 10.1016/0168-583X(93)95803-D.

Harris, J.T., Spacecraft charging at geosynchronous altitudes: Current balance and critical temperature in a non-Maxwellian plasma, Thesis, AFIT/GAP/ENP/03-05, Air Force Institute of Technology, Wright-Patterson AFB, OH, (2003).

Hasselkamp, D., S. Hippler, and A. Scharmann, Ion-induced secondary electron spectra from clean metal surfaces, *Nucl. Inst. Methods Phys. Res.* B, Vol. 18 (1), 561–565, (1986), 10.1016/S0168-583X(86)80088-X.

Katz, I., D.E. Parks, M.J. Mandell, J.M. Harvey, D.H. Brownell, S.S. Wang, and M. Rotenberg, A Three dimensional dynamic study of electrostatic charging in materials, NASA CR-135256, NASA, Washington, D.C., (1977).

Lai, S.T., Critical temperature for the onset of spacecraft charging; The full story, AIAA Plasmadynamics Conf., Seattle, WA, Paper AIAA 2008-3782, (2008).

Lai, S.T., An improved Langmuir probe formula for modeling satellite interactions with near geostationary environment, *J. Geophys. Res.*, Vol. 99 (A1), 459–468, (1994).

Lai, S.T., and M. Tautz, High-level spacecraft charging in eclipse at geosynchronous altitudes: A statistical study, *J. Geophys. Res.*, Vol. 111, A09201, (2006), 10.1029/2004JA010733.

Meyer-Vernet, N., How does the solar wind blow? A simple kinetic model, *Eur. J. Phys.*, Vol. 20 (3), 167–176, (1999).

Mott-Smith, H.M., and I. Langmuir, The theory of collectors in gaseous discharges, *Phys. Rev.*, Vol. 28, 727–763, (1926).

Rajopadhye, N.R., V.A. Joglekar, V.N. Bhoraskar, and S.V. Bhoraskar, Ion secondary electron emission from Al_2O_3 and MgO films, *Solid State Comm.*, Vol. 60(8), 675–679, (1986).

Vasyliunas, V.M., A survey of low-energy electrons in the evening sector of the magnetosphere with Ogo 1 and Ogo 3. *J. Geophys. Res.*, Vol. 73, 2339–2385, (1968).

Whipple, E.C., Jr., Potential of surface in space, *Rep. Prog. Phys.*, Vol. 44, 1197–1250, (1981).

CHAPTER 9

Spacecraft Charging in Sunlight

9.1 THE PHOTOELECTRIC EFFECT

Albert Einstein won his Nobel Prize for his discovery of the photoelectric effect (see, for example, https://en.wikipedia.org/wiki/Photoelectric_effect). With this effect, photons (light particles) impinging on surfaces may generate photoelectrons. This phenomenon brings forth some important questions for spacecraft charging:

1. Can sunlight generate photoelectrons from spacecraft surfaces?
2. Does photoemission depend on surface properties and surface conditions?
3. Does the photoelectron current add a new player to the balance of currents?
4. How does the photoelectron current affect the determination of the spacecraft potential?
5. Is there a consequence from the fact that sunlight shines from one direction only, whereas the ambient electrons come from all directions?
6. Is there a mitigation method for spacecraft charging in sunlight?

DOI: 10.1201/9780429275043-9

9.2 PHOTOELECTRON EMISSION

Light of various wavelengths have different energies. The energy E_{ph} of a photon is given by its frequency ω multiplied by Planck's constant h.

$$E_{ph} = h\omega \tag{9.1}$$

To generate photoelectrons from a surface, it is necessary to have the photon energy exceeding the work function W of the surface material. In this manner, the work function W behaves like a departure tax. The Lyman Alpha line (see Figure 9.1) is the most important ultraviolet spectral line in sunlight and has about 10.2 eV in energy, which exceeds the work function (about 3 to 4 eV) of typical spacecraft surfaces at geosynchronous altitudes. Different materials have different work functions. For a list of work functions of surface materials, see, for example, *CRC Handbook of Chemistry and Physics* [2020].

Consider a surface material with a work function W of about 4 eV. It is often a good approximation to assume that the Lyman Apha line is mainly responsible for generating the photoelectrons [*Kellogg*, 1980], because the visible lines are of lower energies and the higher energy lines are of lower amplitudes. If the work function W = 4 eV, the photoelectron generated by the Lyman Alpha line would come out from the surface with energy, E_{ph}, of about

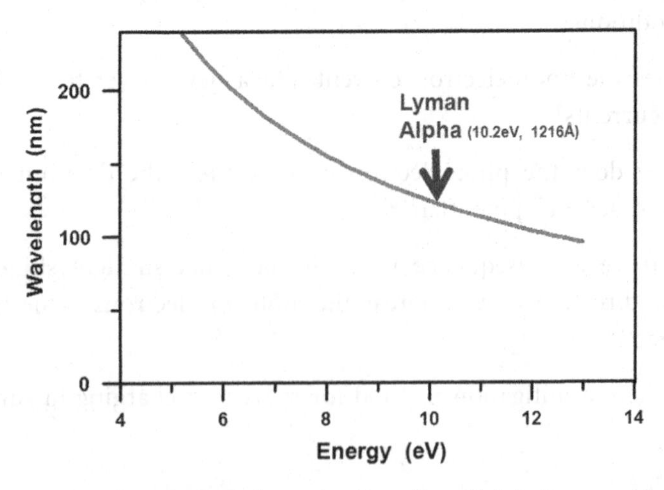

FIGURE 9.1 Wavelength ω and energy E_{ph} of sunlight's Lyman Alpha spectral line.

$$E_{ph} = 10.2 - W \approx 6.2\text{eV} \tag{9.2}$$

If the photoelectron loses some energy by collisions on its way out the material, the photoelectron would have less energy. The energy loss by collisions has to satisfy quantum conditions, which, for our purpose, will be neglected because the loss is small.

9.3 PHOTOELECTRON CURRENT

A spacecraft spends most of its time in sunlight and less time in eclipse. The photoelectron current I_{ph} emitted from a spacecraft of negative potential ϕ is an important current participating in the current balance.

$$I_e(\phi) + I_S(\phi) + I_B(\phi) + I_i(\phi) + I_{Si}(\phi) + I_{ph}(\phi) = 0 \tag{9.3}$$

where the currents are of the incoming ambient electrons $I_e(\phi)$, the outgoing secondary electrons $I_S(\phi)$, the backscattered electrons $I_B(\phi)$, the incoming ambient ions $I_i(\phi)$, the outgoing ion-induced secondary electrons $I_{si}(\phi)$, and the outgoing photoelectrons $I_{ph}(\phi)$. If you can solve this equation, (9.3), you can find out the spacecraft potential ϕ.

The ambient electron flux J_e at geosynchronous altitude is 0.115×10^{-9} Acm^{-2} on the average [*Purvis et al., 1984*]. The outgoing electron flux J_{ph} from typical spacecraft surfaces is 2×10^{-9} Acm^{-2} [*Stannard, et al.*, 1981]. The outgoing photoelectron current J_{ph} exceeds the incoming ambient electron current J_e by about 20 times.

$$J_{ph} \approx J_e \times 20 \tag{9.4}$$

In (9.3), which current is the dominant one? Since current is flux multiplied by the area and the photoelectron flux greatly exceeds that of the ambient electrons, it is obvious that the photoelectron current term $I_{ph}(\phi)$ is the dominant one. However, this fact leads to another obvious question: "If $I_{ph}(\phi)$ is dominant over all the other currents in (9.3), how can current balance occur?" This is because all the other currents added together cannot match the photoelectron current $I_{ph}(\phi)$ in (9.3), which therefore cannot balance to zero [Figure 9.2].

From the above reasoning, one may firmly believe that, since the photoelectron current greatly exceeds all the other currents, spacecraft charging to negative potentials cannot occur in sunlight. Yet, from time to time, spacecraft operators and spacecraft researchers have observed,

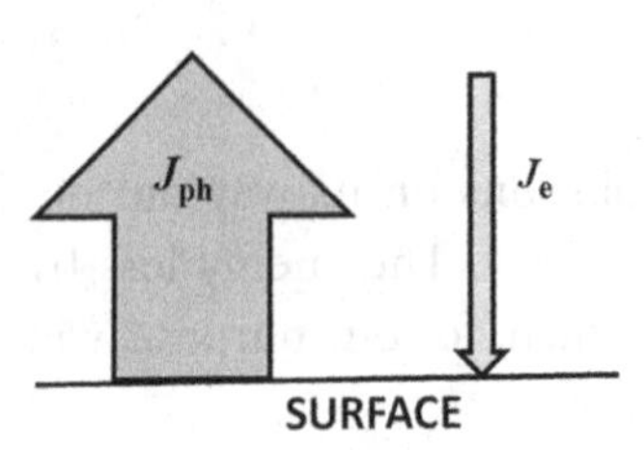

FIGURE 9.2 Competition between the photoelectron flux J_{ph} and the ambient electron flux J_e at geosynchronous altitudes. The outgoing photoelectron flux J_{ph} is about 20 times larger than the incoming ambient electron flux J_e on the average. How can a satellite charge to negative potentials in sunlight?

by means of the instruments onboard, spacecraft charging does occur in sunlight. Why? We will explore for an answer in Section 9.5.

9.4 CHARGING TO POSITIVE POTENTIAL

The energy of a photoelectron generated by sunlight is typically only a few eV. Spacecraft charging by photoelectrons is typically to a few positive volts (+V) only. If the charging level ϕ (+V) exceeds the photoelectron energy E_{ph}, the photoelectron must return and therefore cannot escape.

If there are other participating currents comparable with the photoelectron current emitted from the sunlit surface of a spacecraft, the current balance of all the comparable currents should be considered for the equilibrium potential ϕ. If the photoelectron current is much larger than all the other currents, then the current balance is mainly between the emitted photoelectron current and the returning photoelectron current.

Therefore, for a conducting spacecraft, the charging level in sunlight is typically a few positive volts only. Such a low level is not expected to cause any significant harm to the scientific measurements onboard or to the spacecraft electronics.

9.5 PHOTOELECTRON YIELD

The number of photoelectrons generated per incoming photon of energy E is the photoelectron yield, or PEY, $Y(E_{ph})$. It is a probability. By doing experimental measurements many times, one can obtain a good average of the measurement value. Some examples of measured photoelectron yields $Y(E_{ph})$ generated from different surface materials are shown in Figure 1 of *Feuerbacher and Fitton* [1972].

Note that, for each of the different surface materials [*Feuerbacher and Fitton*, 1972], the PEY $Y(E_{ph})$ starts off from $Y(E_{ph}) = 0$ at just above the work function energy W of the surface material, climbs to a peak, and then decreases slowly.

If the incoming photons are not mono-energetic but form a distribution $f(\omega)$, the photoelectron flux J_{ph} can be written as an integral over all frequencies ω.

$$J_{ph} = \int_0^{\infty} d\omega f(\omega) Y(\omega) \tag{9.5}$$

where $f(\omega)$ is the incoming solar photon flux, and the photoelectron yield $Y(\omega)$ is written as a function of the photon energy ω related to the photon energy E_{ph} by (9.1). The integral in (9.5) may as well start at $\omega = W/h$ because the photons of lower frequency are not energetic enough to generate photoelectrons.

If the solar photons of frequency ω are coming in at various angles θ of incidence, the yield function should also be angle dependent. In that case, (9.5) is written as follows:

$$J_{ph} = \int_{-\pi}^{\pi} d\theta \int_0^{\infty} d\omega \; f(\omega, \theta) Y(\omega, \theta) \tag{9.6}$$

where $f(\omega,\theta)$ is the angle-dependent photon flux and $Y(\omega,\theta)$ the angle-dependent photoelectron yield function.

9.6 SURFACE CONDITION

The reflectance $R(\omega)$ is an important factor for photoelectron yield. If the surface is highly reflective, the photoelectron yield $Y(\omega,\theta)$ is reduced [*Samson*, 1967].

$$Y(\omega, \theta) = Y_0(\omega, \theta) \; [1 - R(\omega, \theta)] \tag{9.7}$$

where $Y_0(\omega,\theta)$ is the photoelectron yield at zero reflectance and $R(\omega,\theta)$ is the reflectance. At zero reflectance, $R(\omega,\theta) = 0$. At perfectly high (100%) reflectance, $R(\omega,\theta) = 1$. Eq(9.7) implies that at perfectly high (100%) reflectance, $R(\omega,\theta) = 1$, and therefore the photoelectron yield $Y(\omega,\theta) = 0$.

$$\lim_{R \to 1} Y(\omega, \theta) \to 0 \tag{9.8}$$

9.6.1 Important Property 1

If the photoelectron yield $Y(\omega,\theta)$ is zero, there is no photoelectron flux J_{ph} emitted from the surface, even though the surface is in sunlight.

Physically, the energy of an incoming photon can be carried away by the outgoing reflected photon. If the reflectance R is unity, or approaching unity, most of the photon energy is carried away, leaving behind almost none, or too little, for paying the departure tax W, in the material for generating photoelectrons.

9.6.2 Important Property 2

If there is no photoelectron current emitted, the surface can charge to negative potentials as if it were in eclipse. With high-energy (keV) ambient electrons coming in, the surface can charge to high potentials (–kV) in sunlight or in eclipse. There are important consequences of this property.

Property 2, as stated above, is a consequence of Property 1. To prove Property 2 in the laboratory, one should establish Property 1 first. Although the physics of Property 1 seems obvious, that of Property 2 is not so obvious [*Lai, 2005*; *Lai and Tautz, 2005, 2006*]. At the moment of this writing, no laboratory experiment to demonstrate Property 2 has been reported in the literature. It remains a research opportunity for graduate students to conduct such an experiment. The results would be important, for they would uphold, refute, or clarify the validity of these properties.

9.7 DIFFERENTIAL CHARGING IN SUNLIGHT

Let us study an example of possible differential charging as a result of different surface reflectance. Suppose two adjacent sunlit surfaces of a geosynchronous satellite are of very different reflectance. Suppose, furthermore, the satellite environment is stormy with abundant energetic (keV) ambient electrons. Both surfaces are receiving significant hot (keV) electron fluxes from the ambient environment. If there were no photoelectron emission, both surfaces would charge to high-level (–kV) negative voltages.

Suppose the first surface gets rid of its electrons received by emitting photoelectrons. It can only do so if the photons can transfer their energies to the surface material, which then emits photoelectrons. If the photoelectron flux J_{ph} from the first surface exceeds the ambient electron flux J_e, there would be no negative surface charging for this surface.

On the other hand, if the second surface has very high reflectance (R near unity), little or no photon energy is transferred to the surface material for generating photoelectrons. As a result, the second surface cannot get

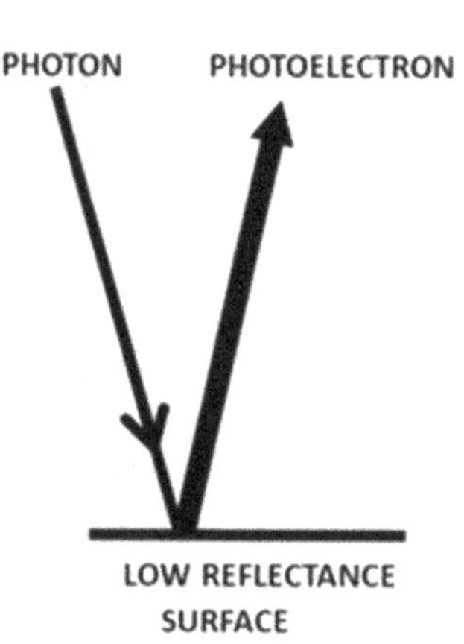

FIGURE 9.3 (Left) Photon reflection from a high reflectance surface and (Right) Photoelectron generation from a low reflectance surface. Their charging behaviors are different in a "hot" ambient plasma environment in space. A high reflectance surface may charge to a high-level voltage in sunlight as if in eclipse. A surface with photoelectron emission may prevent charging to negative voltage.

rid of its surface electrons received from the ambient plasma. Therefore, the second surface may charge to high (–kV) level [Figure 9.3].

9.8 POSSIBLE SCENARIOS OF DIFFERENTIAL CHARGING

9.8.1 Bi-Reflectance Surface Pair

If these two surfaces of very different reflectance are very near each other, their very different potentials may form strong electric fields between the two surfaces. Let us call such a surface pair a bi-reflectance surface pair. It is a differential charging situation. By itself, a discharge between such nearby surfaces of different voltages may or may not happen, but there is a probability that a discharge may happen. If there is something that may trigger a discharge, the probability increases. For example, if there is ionization of neutral gas in the area near, or between, the two surfaces, the probability increases. Or, if there is plasma situated between the surfaces of vastly different voltages, the situation is called a triple junction [*Cho*, 2011], which favors increased probability of discharges.

9.8.2 Eclipse Exit

As another example of a possible scenario, let us consider a pair of bi-reflectance surfaces on a satellite that is entering an eclipse

(Earth's shadow) from the dusk side and later exiting the eclipse to the morning side. The ambient electrons are usually not energetic on the dusk side, so both surfaces do not charge to high negative potentials there. Inside the eclipse, they may experience high energy (keV) ambient electrons, which may travel sunward from the geomagnetic tail. There, both surfaces may charge to high negative voltages without any photoelectron emissions. Upon the eclipse exit, the ambient electron environment may be energetic (keV) in the early hours on the morning side. The "hot" electrons impact on both of the surfaces. There, the surface with high reflectance emits no photoelectrons and therefore charges to a high negative potential in sunlight. At the same time, the other surface, with low reflectance, absorbs the sunlight photon energy and converts it to photoelectron emission. There, this surface does not charge to high negative voltage. So now, we have a differential charging situation between the two surfaces. The situation poses a hazardous risk for discharges or even for electronics operational anomalies. A lesson is learned here: it may be more risky for anomalies to occur after an eclipse exit than before an eclipse entrance [Figure 9.4].

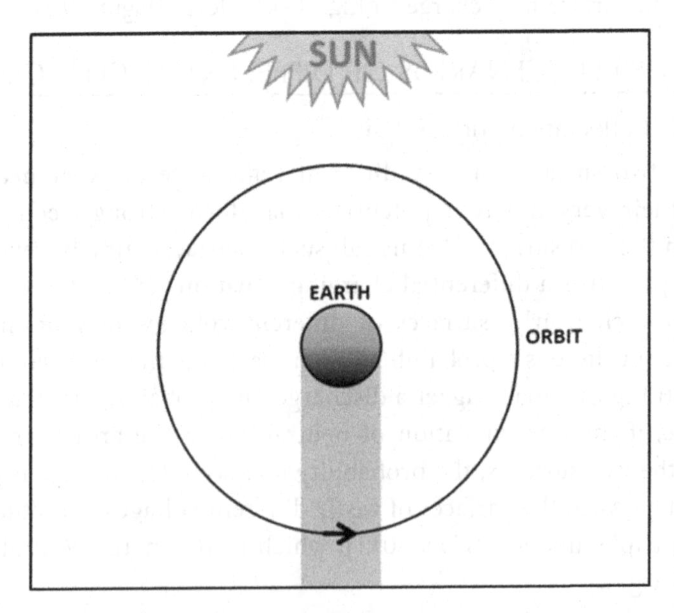

FIGURE 9.4 Eclipse entrance and exit. A satellite enters an eclipse from the dusk side and exits from the dawn side. The ambient electrons are more energetic in the early hours on the morning side (right-hand side). The arrow shows the direction of travel of the satellite.

9.9 SUMMARY

Several salient points in this chapter are summarized as follows.

1. **Photoelectric effect.** Photons of frequency ω impacting on a surface can generate photoelectrons of E_{ph}. The photon energy $h\omega$ is converted to the kinetic energy of the photoelectron. The energy required for a photoelectron to be generated from the surface is at least the work function energy W of the surface material.
2. **Photoelectron energy.** The work function W is analogous to a departure tax. For typical spacecraft surface materials, W is about 3 to 4 eV. The salient ultraviolet spectral line of sunlight is the Lyman Alpha line, which is 10.2 eV. Photoelectrons emitted from spacecraft surfaces have low energies of a few eV.
3. **Photoelectron flux.** The photoelectron flux emitted from spacecraft surfaces is about 2×10^{-9} Acm^{-2}, which is about 20 times larger than the average ambient electron flux 0.115×10^{-9} Acm^{-2} at the geosynchronous space environment.
4. **Spacecraft charging by photoemission.** The photoelectron current adds to the sum of all currents to and from a spacecraft. The balance of all currents governs the spacecraft potential. Since the photoelectron flux exceeds all the fluxes combined, there is no balance. But, the photoelectrons have only a few eV in energy, so they return to the spacecraft as soon as the spacecraft potential reaches a few volts positive. The current balance is mainly between the outgoing and the returning photoelectron currents only. Indeed, spacecraft charging of conducting spacecraft surfaces in sunlight is usually a few positive volts only.
5. **Surface reflectance.** (1) For surfaces with high reflectance, the reflected photon efficiently takes away the energy necessary for photoemission. Therefore, high reflectance of surfaces should have low, or no, photoemission. (2) Highly reflecting surfaces should charge to high negative voltages by the bombardment of "hot" electrons if they are present. Therefore, they are capable of charging to high negative potentials in sunlight as if in eclipse. Laboratory experiments are needed to confirm, refute, or clarify this conjecture.
6. **Differential charging in sunlight.** If two adjacent surfaces with different properties, such as reflectance, charge to very different

potentials, there is a risk that a discharge between the two surfaces may occur. If there is some factor that may trigger a discharge, the risk would be enhanced. "Triple junction" and "eclipse exit" are possible differential charging scenarios that may risk discharges.

EXERCISES

1. Suppose photons of 2 eV in energy are incident on a surface whose work function is 3.5 eV. Can photoelectrons be generated?

2. Suppose photoelectrons of 2 eV in energy are emitted from a surface and the ambient electron flux is much smaller than the photoelectron flux. Can positive voltage charging occur? What is the maximum charging potential?

REFERENCES

Cho, M., Surface discharge on spacecraft, in *Spacecraft Charging*, S.T. Lai (ed.), vol. 237, Progress in Astronautics and Aeronautics, AIAA Press, Reston, VA, USA, (2011).

CRC Handbook of Chemistry and Physics, 101st edition, J. Rumble (ed.), CRC Press, UK, (2020), ISBN 9780367417246.

Feuerbacher, B., and B. Fitton, Experimental investigation of photoemission from satellite surface materials, *J. Appl. Phys.*, Vol. 43 (4), 1563–1572, (1972). https://en.wikipedia.org/wiki/Photoelectric_effect

Kellogg, P. J., Measurements of potential of a cylindrical monopole antenna on a rotating spacecraft, *J. Geophys. Res.*, Vol. 85 (A10), 5157–5159, (1980).

Lai, S.T., Charging of mirror surfaces in space, *J. Geophys. Res.*, Vol. 110 (A1), A01204, (2005), 10.1029/2002JA009447.

Lai, S.T., and M. Tautz, Aspects of spacecraft charging in sunlight, *IEEE Trans. Plasma Sci.*, Vol. 34 (5), 2053–2061, (2006). 10.1109/TPS.2006.883362.

Lai, S.T., and M. Tautz, Why do spacecraft charge in sunlight? Differential charging and surface condition, In Proceedings of the 9th Spacecraft Charging Technology Conference, Paper 9, pp. 10, JAXA-SP-05-001E, Tsukuba, Japan, (2005).

Purvis, C.K., H.B. Garrett, A.C. Whittlesey, and N.J. Stevens, Design Guidelines for Assessing and Controlling Spacecraft Charging Effects, NASA Tech. Paper 2361, NASA, (1984).

Samson, J.A.R. *Techniques of Ultraviolet Spectroscopy*, John Wiley, Hoboken, NJ, (1967).

Stannard, P.R., I. Katz, M.J. Mandell, J.J. Cassidy, D.E. Parks, M. Rotenberg, and P.G. Steen, Analysis of the charging of the SCATHA (P78-20) satellite, NASA CR-165348, (1981).

CHAPTER 10

The Monopole-Dipole Model

10.1 INTRODUCTION

A spacecraft spends most of its time in sunlight. Therefore, charging in sunlight is an important aspect (see Chapter 9). For a spacecraft with all-conducting surfaces, the potential is uniform because the electrons can travel with little resistance on the surfaces. Since the photoelectron flux emitted from the spacecraft surfaces exceeds that of the ambient electrons usually, the potential of an all-conducting spacecraft is usually a few positive volts only [*Kellogg*, 1980; *Lai et al., 1986*].

For a spacecraft with mostly dielectric (nonconducting) surfaces, the potential is not uniform. The ambient electrons are coming in from all directions, whereas sunlight comes in from the direction of the Sun only. Although both the sunlit side and the shadowed side can receive ambient electrons, which may be of high energy (keV), only the sunlit side can emit photoelectrons, while the shadowed side cannot. Since the photoelectron flux J_{ph}, which is outgoing from a sunlit surface, exceeds the ambient electron flux J_e, which is incoming to the surface, by about 20 times on the average (see Chapter 9), the surface is expected to charge to positive potentials.

$$J_{ph} \approx 20 \times J_e \qquad (10.1)$$

DOI: 10.1201/9780429275043-10

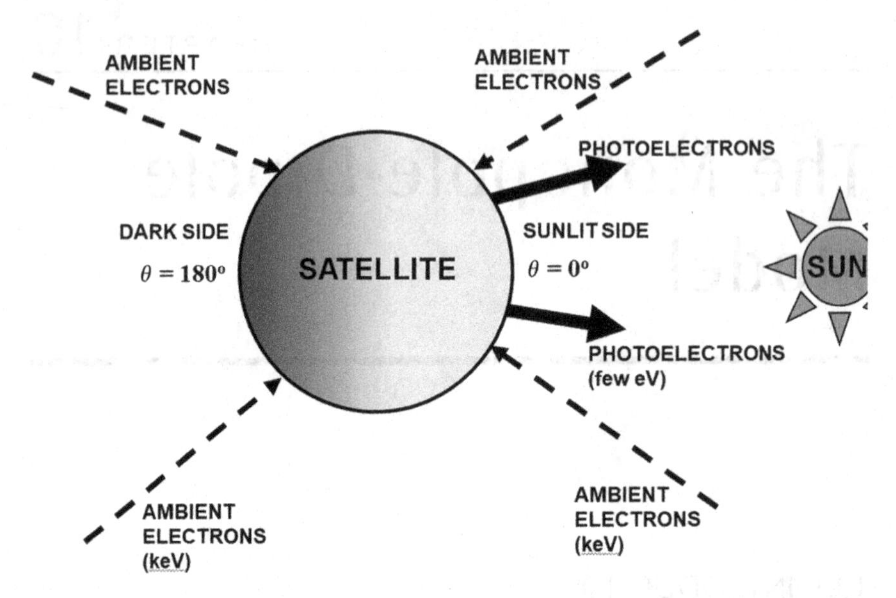

FIGURE 10.1 Incoming ambient electrons and outgoing photoelectrons. The ambient electrons are from all directions, but the outgoing photoelectrons are from the sunlit surfaces only.

Since the shadowed side receives no sunlight but only ambient electrons, the shadowed side, emitting no photoelectrons, can charge to high-level (−kV) negative potentials. The result is a non-uniform spacecraft charging distribution, one side charging to low positive potentials while, at the same time, the other side can charge to high-level negative potentials [Figure 10.1].

As a result of the high negative potential compared with the low positive potential (a few volts) on the sunlit side, the high-level negative potential contours of the shadowed side can wrap to the sunlit side. If the wrapping is complete, the negative potentials can block the escape of the photoelectrons, which have a few eV in energy only. As a result, the entire satellite may charge to negative potentials everywhere.

10.2 THE MONOPOLE-DIPOLE MODEL

The monopole-dipole model [*Besse and Rubin*, 1980] describes such a charge distribution approximately. It is useful as a handy tool for physical description and for analytical calculations. It is as follows:

$$\phi(\theta, r) = K\left(\frac{1}{r} - \frac{A\cos\theta}{r^2}\right) \tag{10.2}$$

where ϕ is the potential at the point (θ, r). The sun angle $\theta = 0°$ is the sun direction. The radial distance r is from the center of the satellite, K the monopole strength, and A the normalized dipole strength. The first term on the right-hand side is the monopole term and the second term the dipole term.

An example of the potential contours obtained from eq(10.2) is shown in Figure 10.2. The parameter A used is $A = 3/4$. The contours from the left-hand side are −50, −70, −80, −90 V, and so on. The potential barrier forms a saddle point located at $(r = R_S, \theta = 0°)$. In the figure, the saddle point is at the meridian $(\theta = 0°)$ of the −80 V contour crossing itself. The surface potentials at $\theta = 0°$ and $\theta = 180°$ are −60 V and −420 V, respectively.

The radial distance R_S of the potential barrier at any θ can be obtained as follows.

$$\left[\frac{d\phi(\theta, r)}{dr}\right]_{r=R_S} = 0 \tag{10.3}$$

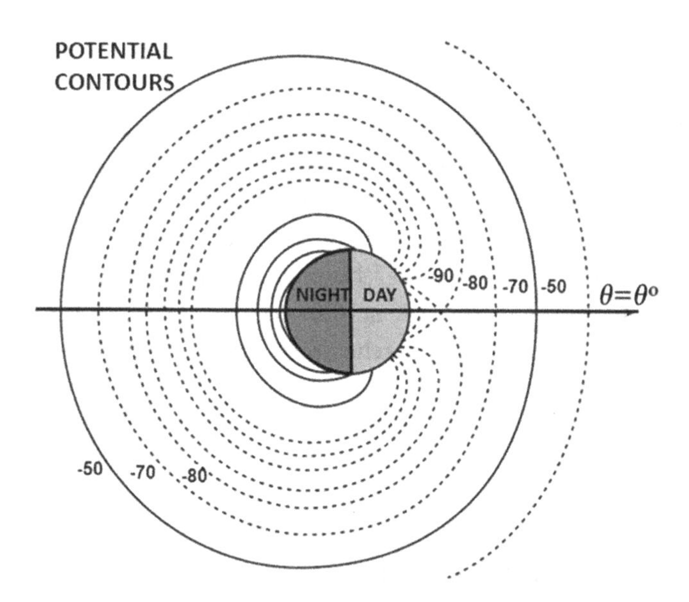

FIGURE 10.2 Potential contours of monopole-dipole model. The parameter A used is 3/4. The contours from left are −50, −70, −80, −90, −100, −110 V, and so on. The saddle point is where the −80 V contour crosses itself at $\theta = 0°$ facing the sun. The surface potentials at $\theta = 0°$ and $\theta = 180°$ are −60 V and −420 V, respectively.

Source: Adapted from Besse and Rubin, 1980. AGU Permission.

which yields

$$R_S = 2A \cos\theta \tag{10.4}$$

Since the potential barrier must be outside the satellite, we obtain from eq(10.4) an inequality:

$$2A \cos\theta > R \tag{10.5}$$

where R is the satellite radius. Since, in a Taylor series of multipole expansion, the monopole term must be greater than the dipole term, eq(10.5) together with eq(10.2) give the following inequality:

$$R > A \cos\theta > R/2 \tag{10.6}$$

The magnitude of the potential barrier $B(\theta)$ at the sun angle θ is given by Lai and Cahoy, [2015].

$$B(\theta) = \phi(\theta, R_S) - \phi(\phi, R) = K\frac{(2A \cos\theta - R)^2}{4A \cos\theta R^2} \tag{10.7}$$

Since photoelectrons have only a few electron volts in energy, a potential barrier $B(\theta)$ of a few electron volts or a little more is sufficient to block the photoelectrons. Therefore, the ratio $B(\theta)/K$ of the barrier to the monopole potentials in eq(10.7) need to be very small in order to be sufficient for blocking, and thereby trapping partially or totally, the photoelectrons emitted from the sunlit surface at (R,θ). In the limit that the potential barrier $B(\theta)$ approaches zero, eq(10.7) gives

$$\lim_{B\to 0} A \cos\theta = R/2 \tag{10.8}$$

10.3 ILLUSTRATIVE EXAMPLES

For illustration of the monopole-dipole model, Figure 10.3 shows the potential profiles along the radial distance r on the sunlit side ($\theta = 0°$) and the dark side ($\theta = 180°$). The satellite radius is 1, the monopole strength is $K = -1$, and the sun is on the right-hand side.

In Figure 10.3, the potential profiles are normalized by the magnitude of the potential $\phi(1, 180°)$. The A parameters used are $A = 0.5, 0.7, 0.9, 1.1$, and

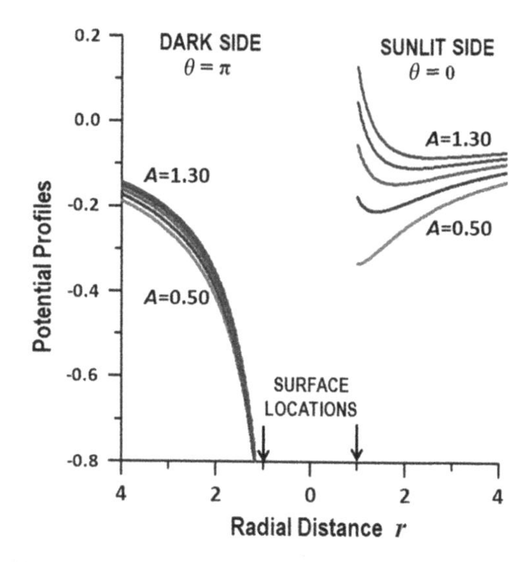

FIGURE 10.3 Monopole-dipole potential profiles along the radial distance at $\theta = 0°$ and $\theta = \pi°$. The satellite radius = 1, the sun is on the right-hand side, and the monopole strength $K = -1$. The potentials are normalized by the potential $\phi(1, \pi°)$. Note that a potential barrier (a dip) forms on the sunlit side when $A > 0.5$.

1.3. Note that a potential barrier (a dip) forms on the sunlit side ($\theta = 0°$) when $A > 0.5$ (see eq.10.5). For A less than 0.5, there is no potential well, and therefore the photoelectrons completely escape.

Figure 10.4 shows the potential profiles along the radial distance at $\theta = 45°$. The satellite radius and monopole strength are the same as in the previous figure. The A parameters are labeled as 0.5, 0.75, 1.0, 1.25, and 1.5. The potentials are normalized in the same way as before. Note that the top three profiles have potential barriers (dips for negative potentials) but the lower ones have not. Eq(10.5) requires that A has to satisfy the inequality $A > (R/2)/\cos(48°)$.

10.4 FRACTION OF ELECTRON FLUX ESCAPING

Suppose the low energy electrons, consisting of photoelectrons and secondary electrons, are trapped, totally or partially, by the potential barrier B, so that they bounce and scatter with each other forming an equilibrium distribution $f(E)$. The fraction L of the electron flux escaping over the barrier can be calculated as

$$L = \frac{\int_B^\infty dE \;\; E \;\; f(E)}{\int_0^\infty dE \;\; E \;\; f(E)} \tag{10.9}$$

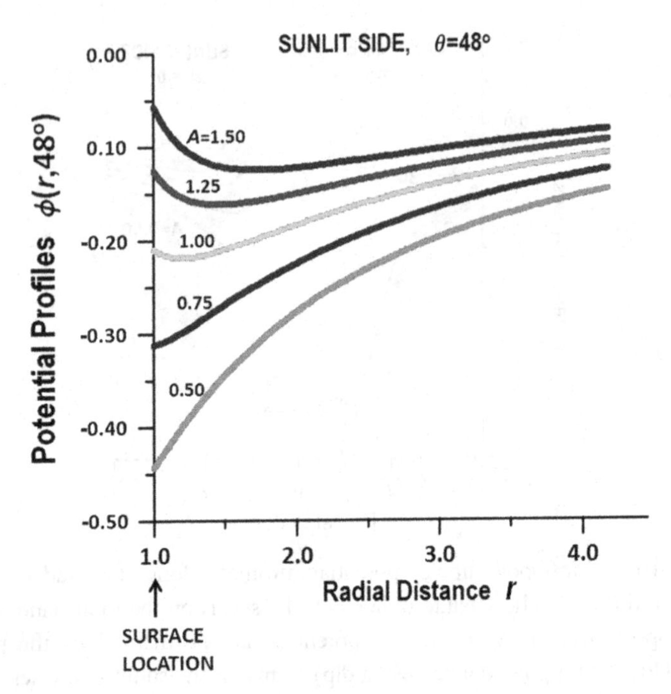

FIGURE 10.4 Potential profiles as functions of radial distancerat sun angle 48°. The satellite surface is at the origin $r = 1$ on the x-axis shown. The monopole strength is $K = -1$. The top three potential profiles, with $A = 1.5$, 1.25, and 1.0, have potential barriers, but the lower two profiles with $A = 0.75$ and 0.5 do not have barriers.

where we have used $E = (1/2)mv^2$. It is often a good approximation to assume that the distribution is Maxwellian. As a result, eq(10.9) can be calculated analytically.

$$L = \frac{\int_B^\infty dE \; E\exp(-E/kT)}{\int_0^\infty dE \; E \; \exp(-E/kT)} = \left(\frac{B}{kT} + 1\right)\exp\left(-\frac{B}{kT}\right) \tag{10.10}$$

where T is the temperature of the photoelectrons and secondary electrons and equals 1.5 eV [*Kellogg*, 1980; *Lai et al., 1986*] approximately. The fraction L in eq(10.10) is a function of the barrier B and the kinetic energy E of the trapped photoelectrons and secondary electrons. Physically, if the barrier B increases and the temperature T is low, L would decrease implying that less electrons would escape.

10.5 EVIDENCE OF TRAPPED LOW-ENERGY ELECTRONS

Figure 10.5 shows the low-energy (below 20 eV) photoelectrons and secondary electrons measured on the Los Alamos National Laboratory (LANL) Satellite 1994-084 at geosynchronous altitudes on the morning side of March 8, 1999. The x-axis shows the time. On the dusk side (left of the eclipse), there was no significant spacecraft charging, because the ambient electron energy was low. There is a blue band from about 16.8 to 17.9 UT(Hours). The arrow below 17 UT indicates local time midnight. The blue band indicates eclipse. No low-energy electrons were in the band because there was no photoelectron, and low-energy ambient electrons were repelled by the charging in eclipse, which was caused by the incoming high-energy ambient electrons.

After eclipse exit, the ambient morning electrons were much more energetic than those in the dusk sector. As a result, high-level negative charging occurred after exit in the morning hours. The satellite was not completely conducting. Monopole-dipole charging occurred there, resulting in the enhanced electron density (red) of about 10 to 12 eV or less in the morning period from 18 UT onward. The occurrence

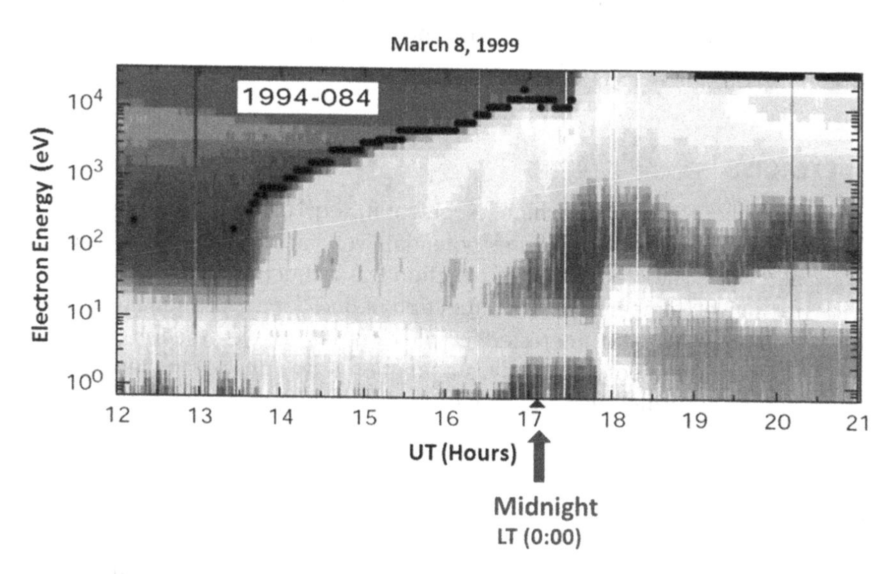

FIGURE 10.5 Evidence of trapped photoelectrons and secondary electrons. Low-energy (below 20 eV) electrons trapped on the sunlit side of the Los Angeles National Laboratory (LANL) Satellite 1994-084 on the morning side of March 8, 1999.

Source: From Thomsen, et al. [2002]. AGU permission.

of trapped low-energy electrons requires the presence of high-level negative charging on the shadowed side and photoemission on the sunlit side. The high-level negative potential contours wrap to the sunlit side and block the escape of the photoelectrons, which are of low-energy. As a result, the entire satellite can become negatively charged. Secondary electrons can be emitted, but they are of the about the same energy as the photoelectrons, and therefore they are blocked together with the photoelectrons.

10.6 "ONE-THIRD" POTENTIAL RATIO

Eq(10.2) of the monopole-dipole model gives the ratio of the front-to-back potentials on the spacecraft surface as follows:

$$\frac{\phi(0°, 1)}{\phi(180°, 1)} = \frac{1 - A}{1 + A} = \frac{1 - 1/2}{1 + 1/2} = \frac{1}{3} \tag{10.11}$$

which is neat and simple. It means that the potential $\phi(\theta = 0°, r = 1)$ on the sunlit side equals one-third of the potential $\phi(\theta = 180°, r = 1)$ on the shadowed side. The one-third potentials ratio, (10.11), is valid if the conditions for the monopole-dipole model are satisfied. The conditions are the inequalities eq(10.5) and eq(10.6).

EXERCISE

1. Plot the fraction L, which is given in eq(10.10), of low-energy electrons escaping from the potential well facing the saddle point at $\theta = 0°$. The plot can be the fraction L as a function of the barrier B for a given temperature T or a function of B/kT for a given sun angle θ. You may also be interested in plotting 1-L as the fraction being trapped in the potential well as a function of various parameters. Or, instead of using the Maxwellian distribution function, you may use other models.

REFERENCES

Besse, A.L., and A.G. Rubin, A simple analysis of spacecraft charging involving blocked photoelectron currents, *J. Geophys. Res.*, Vol. 85 (5), 2324–2328, (1980).

Kellogg, P.J., Measurements of potential of a cylindrical monopole antenna on a rotating spacecraft, *J. Geophys. Res.*, Vol. 85 (10), 5157, (1980).

Lai, S.T., H.A. Cohen, T.L. Aggson, and W.J. McNeil, Boom potential on a rotating satellite in sunlight, *J. Geophys. Res.*, Vol. 91, 12137–12141, (1986).

Lai, S.T., and K. Cahoy, Trapping of photoelectrons during spacecraft charging in sunlight, *IEEE Trans. Plasma Sci.*, Vol. 43, No. 9, pp. 2856–2860, DOI: 10.1109/TPS.2015.2453370, (2015).

Thomsen, M.F., H. Korth, and R.C. Elphic, Upper cutoff energy of the electron plasma sheet as a measure of magnetospheric convection strength, *J. Geophys. Res.*, Vol. 107 (A10), 1331, (2002).

CHAPTER 11

The Question of Independence on Ambient Electron Density in Spacecraft Charging

11.1 INTRODUCTION

We assume that the reader who has come so far from the early chapters to this one is now familiar with some essential concepts of spacecraft charging. The essential concepts include floating potential, current balance, ambient electron and ion fluxes, secondary and backscattered electron currents, photoelectron currents, and the monopole-dipole model. In this chapter, we will use some of the above concepts together. In this way, this chapter is at a higher level than the previous ones.

It would not be surprising if a space scientist says that his/her spacecraft charges when the ambient electron density is high, whereas another scientist says this is not so. The dilemma reminds us of the ancient fable about four blind persons, each having a firm personal opinion of the shape of an elephant. It is necessary to understand the

DOI: 10.1201/9780429275043-11

situation for the appropriate physical processes to occur. This chapter will explain aspects of spacecraft charging in various situations.

11.2 ONSET OF CHARGING IN MAXWELLIAN PLASMA

Consider the Maxwellian plasma situation in which the onset of spacecraft charging is governed by the current balance between the incoming electrons and the sum of the outgoing electrons plus the incoming ions. Since the ion flux is most likely less than the electron flux by nearly two orders of magnitude, let us ignore the ion current as a good approximation. Thus, we write the current balance equation, eq(6.1), as follows.

$$\int_0^\infty dE \;\; Ef(E) = \int_0^\infty dE \;\; Ef(E)[\delta(E) + \eta(E)] \tag{11.1}$$

The Maxwellian distribution function *f*(*E*), eq(6.4), is as follows.

$$f(E) = n\left(\frac{m}{2\pi kT}\right)^{3/2} \exp\left(-\frac{E}{kT}\right) \tag{11.2}$$

The function *f*(*E*) has a multiplicative factor *n*. The SEY $\delta(E)$ and BEY $\eta(E)$ functions are functions of the primary electron energy *E*, but not the ambient electron density *n*. The incidence angle dependence of $\delta(E)$ and $\eta(E)$ can be neglected without loss of generality here.

Mathematically, the density *n* on both sides of eq(11.1) are cancelled out. This cancellation property means that, in the Maxwellian plasma situation, the onset of spacecraft charging is independent of the ambient electron density *n*. Physically, the cancellation property means that the more ambient electrons are coming in, the more secondary and backscattered electrons are going out proportionally. Therefore, there is no ambient electron density dependence for the onset of spacecraft charging governed by the balance of currents of the incoming ambient electrons and the outgoing secondary and backscattered electrons in the Maxwellian plasma (eq.11.1).

11.3 CHARGING TO FINITE POTENTIALS IN A MAXWELLIAN PLASMA

Negative spacecraft potentials repel the incoming ambient electrons and attract the ambient ions toward the spacecraft. The repulsion of the incoming ambient electrons is by means of the Boltzmann factor. As described in Chapter 5, the secondary and backscattered electrons are outgoing from the spacecraft. The ion attraction is by means of the Langmuir orbit-limiting factor, as described in Chapter 3. The current balance equation for this simple model of spacecraft charging is as follows:

$$I_e(0)\,[1 - <\delta + \eta>]\exp\left(-\frac{q_e\phi}{kT_e}\right) - I_i(0)\left(1 - \frac{q_i\phi}{kT_i}\right) = 0 \qquad (11.3)$$

In eq(11.3), the spacecraft potential ϕ is considered negative. The ambient electrons and ion currents are $I_e(\phi)$ and $I_i(\phi)$, respectively, T_e and T_i are the electron and ion temperatures, and k is the Boltzmann constant. The last parenthesis in eq(11.3) is the Langmuir orbit-limiting factor. It is greater than unity because the ion charge q_i is positive and the spacecraft potential ϕ is negative. The ratio of the outgoing secondary and backscattered electron current to that of the incoming ambient electrons is given by

$$< \delta + \eta > = \frac{\int_0^\infty dEE\ f_e(E)\,[\delta(E) + \eta(E)]}{\int_0^\infty dEE\ f_e(E)} \qquad (11.4)$$

The only electron density dependent terms in eq(11.3) are the electron and ion currents $I_e(0)$ and $I_i(0)$, respectively. The electron and ion currents are functions of their distributions $f_e(E)$ and $f_i(E)$, respectively. The electron and ion densities n_e and n_i are multiplicative factors in $f_e(E)$ and $f_i(E)$, respectively. For neutral space plasmas, the electron and ion densities are equal. Therefore, the electron and ion densities cancel each other in eq(11.3).

11.4 SPACECRAFT CHARGING IN KAPPA PLASMA

In Section 11.2 and Section 11.3, we have argued that spacecraft charging does not depend on the ambient electron density. Both (a) the onset of spacecraft charging and (b) spacecraft charging to finite potentials do not depend on the ambient electron density. The argument for (a) is built on four properties: (1) the only density-dependent terms are the distribution functions, (2) the density term is a multiplicative constant in the Maxwellian distribution, (3) the SEY and BEY yields are not density dependent, and (4) the ambient electron flux exceeds that of the ambient ions by two orders of magnitude. The argument for (b) depends on the first three properties above, and (4) the ambient space plasma is, by and large, neutral.

These arguments are applicable to other distributions as long as the density term is a multiplicative constant. The Kappa plasma distribution [*Vasyliunas*, 1968; *Meyer-Vernet*, 1999], which is often useful, has this property. Therefore, both behaviors, (a) the onset of charging and (b) charging to finite potentials, in a Kappa plasma distribution do not depend on the ambient electron density.

11.5 CHARGING OF CONDUCTING SPACECRAFT IN SUNLIGHT

Surfaces of conducting spacecraft usually charge to a few positive volts in sunlight [Kellogg, 1980; *Lai et al.*, 1986]. This is because the photoelectron flux from surfaces usually exceeds the incoming ambient electron flux by an order of magnitude. According to Purvis et al. [1984], the average ambient flux measured on the SCATHA satellite was $J_e = 0.115 \times 10^{-9}$ Acm^{-2}, whereas *Feuerbacher and Fitton* [1972] reported that the photoelectron flux from typical satellite surfaces was $J_{ph} = 2 \times 10^{-9}$ A cm^{-2}.

For a spacecraft charged to a positive potential $\phi > 0$ in sunlight as a result of the dominating current I_{ph} of photoelectrons leaving the spacecraft, a simple current balance equation is as follows:

$$I_e(0)\left(1 - \frac{q_e\phi}{k\,T_e}\right) = I_i(0)\exp\left(-\frac{q_i\phi}{kT_i}\right) + I_{ph} \tag{11.5}$$

In the above equation, we have neglected the secondary electrons because the spacecraft potential ϕ is positive. Although the ambient electron and ion currents, $I_e(0)$ and $I_i(0)$, are proportional to the ambient electron and ion densities, n_e and n_i, the photoelectron current I_{ph} is not. Therefore, the densities in eq(11.5) do not cancel each other. As a result, the potential ϕ in eq(11.5) depends on the ambient plasma densities. Indeed, measurements on satellites confirmed [*Knott et al.*, 1984; *Escoubet, et al.*, 1997] that the low-level positive spacecraft potential ϕ in sunlight does vary with the ambient electron densities.

11.6 NEGATIVE-VOLTAGE CHARGING OF A CONDUCTING SPACECRAFT IN SUNLIGHT

Normally, charging in sunlight is positive in voltage, because the photoelectron flux typically exceeds the ambient electron flux by an order of magnitude. However, there are three factors for the ambient electron current to exceed the photoelectron current [Lai and Tautz, 2005, 2006].

1. During stormy events in geospace, the ambient electron flux may surge by a few times above normal. For example, *Gussenhoven and Mullen* [1982] observed on the SCATHA satellite a worst-case electron flux of $J_w = 0.501 \times 10^{-9}$ A cm^{-2}, which is nearly five times above the average value. It is possible that higher electron fluxes in extreme stormy events have been observed by other satellites since then. In the simplest geometry, i.e. a sphere, half of the surface area is in sunlight but the entire surface area receives ambient electrons. Therefore, for a sphere, the ambient current derived from $2 \times J_w$ ($=1.1 \times 10^{-9}$ Acm^{-2}) is no longer an order of magnitude less than J_{ph} ($=2 \times 10^{-9}$ A cm^{-2}). If the ratio of the incoming ambient electron flux to the sum of the incoming ion flux plus the outgoing photoelectron flux exceeds unity, which seems rare, then this ratio alone can account for negative voltage charging in sunlight.

2. If the spacecraft geometry features large surface areas in shadows, or even shadows in shadows, the ambient electron current received by all surfaces on the satellite can be much larger than the photoelectron current generated from the smaller sunlit areas only. This factor depends on the artificial design of the satellite.

3. If the surfaces are highly reflective, the photons are reflected with much of their original energy. As a result, the total energy transferred to the surface material may be greatly reduced. To generate a photoelectron, the energy required is the sum of the work function of the surface material plus the kinetic energy of the photoelectron. There may also be some attenuation loss in the surface material. We conjecture that, depending on the reflectance, a highly reflective surface may charge in sunlight as if it were in eclipse [*Lai*, 2005]. This conjecture may serve as a simple,

but important, research topic for the students in the laboratory and in space.

In view of the above factors (1–3), individually or in combination, it is possible that a spacecraft with conducting surfaces can charge to negative voltages even in sunlight. In such an event, the current balance is expected to resemble that of eqs(11.1–11.3), which have been discussed in Section 11.2 and Section 11.3. If so, the onset of charging as well as charging to finite negative voltages should depend on the ambient electron temperature only but not on the ambient electron densities [*Lai, et al.*, 2017].

11.7 CHARGING IN THE MONOPOLE-DIPOLE MODEL

For spacecraft covered by mostly dielectric surfaces, which are not conductors but are useful for protecting sensitive electronic instruments and wires from ambient electrons, the potential distribution around the spacecraft in sunlight is not uniform. The shadowed side, emitting no photoelectrons but receiving ambient electrons, can charge to high negative potential (Chapter 10). The high negative potential can wrap to the sunlit side and may block the photoelectrons from escaping [*Besse and Rubin*, 1980; *Lai and Cahoy*, 2015]. In this manner, the charging of the satellite depends on the high negative voltage charging by the ambient electrons and ambient ions impacting on the shadowed side. High negative voltage charging of the shadowed side by the ambient electrons does not depend on the ambient electron density, because photoelectrons are not emitted there.

11.8 DOUBLE MAXWELLIAN DISTRIBUTION

The space plasma environment may sometimes be better described by a double Maxwellian distribution rather than a single one. A double Maxwellian distribution is a sum of two single Maxwellian distributions, each with its own density and temperature.

$$f(E) = f_1(E) + f_2(E) \tag{11.6}$$

where the functions $f_1(E)$ and $f_2(E)$ are given as in eq(11.2), but the density n and temperature T are replaced by n_1, T_1 and n_2, T_2, respectively.

To determine the equilibrium spacecraft potential ϕ, one substitutes the above $f(E)$ of eq(11.6) into the current balance equations, eq(11.1) to eq(11.4). Since there are two densities, n_1 and n_2, and two temperatures, T_1 and T_2, involved in the current balance equations corresponding to eq(11.1) and eq(11.3), the densities and temperatures cannot be cancelled. Therefore, charging in a double Maxwellian environment does depend on the two ambient electron densities and the two ambient electron temperatures [*Lai*, 1991].

11.9 CHARGING IN THE IONOSPHERE

The net current I of negative charge received by a satellite in the ionosphere is given by [*Mozer*, 1973]:

$$I = An_e\left(\frac{kT_e}{2\pi m_e}\right)^{1/2}\exp\left(-\frac{q_e\phi}{kT_e}\right) - I_p - An_i\left(\frac{kT_i}{2\pi m_i}\right)^{1/2} \tag{11.7}$$

where the first term on the right side is the current of the incoming ambient electrons and the last term is that of the incoming ambient ions. The surface area is A, and the potential ϕ is negative. The term I_p is the current contributed by nonthermal plasma particles. The current balance is given by the zero net current I.

$$An_e\left(\frac{kT_e}{2\pi m_e}\right)^{1/2}\exp\left(-\frac{q_e\phi}{kT_e}\right) - I_p - An_i\left(\frac{kT_i}{2\pi m_i}\right)^{1/2} = 0 \tag{11.8}$$

Eq(11.8) is similar to eq(11.3) with a difference: the I_p term. If this I_p term is negligible relative to the ambient currents, the conclusion of Section 11.3 would hold, at least approximately. For example, if the nonthermal current is solely due to secondary and backscattered electrons, its contribution would be small. This is because at the energies of ionospheric electrons, the yields of secondary and backscattered electrons are small.

If I_p is mainly due to the ram current given by the relative velocity of the spacecraft relative to the ambient plasma, the ram current is proportional to the ambient plasma density. In neutral plasma, the densities in the current balance eq(11.8) would cancel each other. So, in this approximation, the charging of the spacecraft is independent of the ambient electron density.

If, however, the satellite is in the sunlit ionosphere, the situation is similar to Section 11.5. The current balance equation, eq(11.5), accounts for the photoelectron current I_{ph} emitted from the satellite. So, the spacecraft potential ϕ depends on the competition between the current of the outgoing photoelectrons and that of the incoming ambient electrons. Therefore, in this case, the ambient electron current plays a role in controlling the spacecraft potential.

If the sunlit satellite is in the monopole-dipole potential configuration, then Section 11.7 applies. However, the formation of monopole-dipole potential configuration requires high negative voltage on the shadowed side (Chapter 10). In the ionosphere, the ambient electrons are of energies well below 1 eV, so high-level charging of the shadowed side is ruled out. Therefore, we need not consider the monopole-dipole potential situation in the ionosphere.

11.10 FACETS OF SPACECRAFT CHARGING: CRITICAL TEMPERATURE AND THE QUESTION OF DEPENDENCE OF AMBIENT ELECTRON DENSITY

In this chapter, we have discussed the question of dependence of ambient electron density in spacecraft charging [*Lai, et al.*, 2017]. The discussion involves many space environmental situations, such as charging in sunlight, charging in double Maxwellian environment, kappa environment, monopole-dipole charging model of spacecraft with dielectric surfaces in sunlight, etc. A similar discussion involving many space environmental situations has been reported in *Lai* [2020]. The combined results of both *Lai, et al.* [2017] and *Lai* [2020] are summarized as follows in Table 11.1.

TABLE 11.1 Facets of Spacecraft Charging

Space Environment Situation	Critical Temperature	Electron Density
Maxwellian	T^*	Independent
Double Maxwellian	No	Dependent
Kappa	T^*	Independent
Cutoff Maxwellian	T^*	Independent
Monopole-Dipole Model in Sunlight	T^*	Independent
Low-Level Charging in Sunlight	No	Dependent
Plasma Probe on Spacecraft	No	Dependent
Charged Particle Beam Emission	No	Dependent

EXERCISES

Students who have reached this point are more advanced than at the beginning. They should be able to do the following exercises.

1. Work out the problem of critical temperature in a Maxwellian plasma environment.
2. Work out the problem of critical temperature in a Kappa plasma environment.
3. Work out the critical temperature problem in a cut-off plasma distribution environment.
4. Work out the problem of parametric domains of charging, non-charging, and triple-roots in a double Maxwellian plasma environment.
5. Work out the problem of charging contours in a monopole-dipole model. Derive the necessary condition for the saddle point barrier to exist.
6. In situations where the outgoing current of secondary and backscattered electrons is proportional to the incoming current, charging is independent of the ambient electron current. Under what conditions is charging dependent on ambient electron current?

REFERENCES

Besse, A.L., and A.G. Rubin, A simple analysis of spacecraft charging involving blocked photoelectron currents, *J. Geophys. Res.*, Vol. 85 (5), 2324–2328, (1980).

Escoubet, C.P., A. Pedersen, R. Schmidt, and P.A. Lindquist, *J. Geophys. Res.*, Vol. 102, 17595–17609, (1997), 10.1029/97/JA00290

Feuerbacher, B., and B. Fitton, Experimental investigation of photoemission from satellite surface materials, *J. Appl. Phys.*, Vol. 43 (4), 1563–1572, (1972).

Gussenhoven, M.G., and E.G. Mullen, A 'worst case' spacecraft charging environment as observed on SCATHA on 24 April 1979, AIAA Paper 82-0271, (1982).

Kellogg, P.J., Measurements of potential of a cylindrical monopole antenna on a rotating spacecraft, *J. Geophys. Res.*, Vol. 85(A10), 5157–5161, (1980), 10.1029/JA085iA10p05157

Knott, K., P. Décréau, A. Korth, A. Pedersen, and G. Wrenn, The potential of an electrostatically clean geostationary satellite and its use in plasma diagnostics, *Planet. Space Sci.*, Vol. 32(2), 227–237, (1984).

Lai, S.T., Charging of mirror surfaces in space, *J. Geophys. Res.*, Vol. 110 (A1), A01204, (2005), 10.1029/2002JA009447.

Lai, S.T., Facets of spacecraft charging; Critical temperature and dependence on ambient electron density, in ECLOUD'18 Proceedings of the Joint INFN-CERN-ARIES Workshop on Electron-Cloud Effects, 3–7 June 2018, La Biodola, Isola d'Elba, Italy, R. Cimino, G. Rumolo, and F. Zimmermann (Eds.), CERN-2020-007, CERN, Geneva, 137–141, (2020), 10.23732/CYRCP-2020-007.

Lai, S.T., Theory and observation of triple-root jump in spacecraft charging, *J. Geophys. Res.*, Vol. 96 (A11), 19269–19282, (1991).

Lai, S.T., and K. Cahoy, Trapped photoelectrons during spacecraft charging in sunlight, *IEEE Trans. Plasma Sci.*, Vol. 43 (9), 2856–2860, (2015), 10.1109/TPS.2015.2453370.

Lai, S.T., H.A. Cohen, T. L. Aggson, and W. J. McNeil, Boom potential of a rotating satellite in sunlight, *J. Geophys. Res.*, Vol. 91 (A11), 12137–12141, (1986).

Lai, S.T., M. Martinez-Sanchez, K. Cahoy, M. Thomsen, Y. Shprits, W. Lohmeyer, and F. Wong, Does spacecraft potential depend on the ambient electron density?, *IEEE Trans. Plasma Sci.*, Vol. 45 (10), 2875–2884, (2017), 10.1109/TPS.2017.2751002.

Lai, S.T., and M.F. Tautz, Aspects of spacecraft charging in sunlight, *IEEE Trans. Plasma Sci.*, Vol. 34 (5), 2053–2061, (2006).

Lai, S.T., and M. Tautz, Why do spacecraft charge in sunlight? Differential charging and surface condition, In Proceedings of the 9th Spacecraft Charging Technology Conference, Paper 9, Tsukuba, Japan, (2005).

Meyer-Vernet, N., How does the solar wind blow? A simple kinetic model, *Eur. J. Phys.*, Vol. 20, 167–176, (1999), 10.1088/0143-0807/20/3/006

Mozer, F.S., Analysis of techniques for measuring DC and AC electric fields in the magnetosphere, *Space Sci. Rev.*, Vol. 14 (2), 272–313, (1973).

Purvis, C., H.B. Garrett, A.C. Whittlesey, & N.J. Stevens, Design guidelines for assessing and controlling spacecraft charging effects, Tech. Pap. 2361, 44 pp., NASA, Washington, D.C., (1984).

Vasyliunas, V.M., A survey of low-energy electrons in the evening sector of the magnetosphere with OGO 1 and OGO 3, *J. Geophys. Res.*, Vol. 73(9), 2839–2884, (1968), 10.1029/JA073i009p02839

CHAPTER 12

Spacecraft Charging Induced by Beam Emissions

12.1 BRIEF REVIEW

Salient aspects of spacecraft charging without beam emission have been given in the first few chapters. This section briefly reviews them before we introduce beam emissions.

The ambient ion flux is about two orders of magnitude smaller than that of the ambient electrons, because of the mass difference. At geosynchronous altitudes, the ambient plasma is of low density (typically below a few electrons per cm^3) and even lower density during stormy times. The ambient electron temperature is typically below 100 eV but goes up to the keV range during stormy periods. The spacecraft potential at equilibrium is governed by the balance of currents. It takes typically milliseconds to reach equilibrium for most spacecraft.

The current balance equation for a negatively charged spacecraft as controlled by the ambient electron and ion currents is given by

$$I_e(0)[1 - \langle \delta + \eta \rangle]\exp\left(-\frac{q_e \phi}{k T_e}\right) - I_i(0)\left(1 - \frac{q_i \phi}{k T_i}\right)^{\alpha} = 0 \quad (12.1)$$

where the spacecraft potential ϕ is negative, $q_e = -q$, $q_i = +q$, and q is the elementary charge. The exponent α is unity for a sphere, ½ for a long

DOI: 10.1201/9780429275043-12

cylinder, and 0 for a plane. $I_e(\phi)$ and $I_i(\phi)$ are the ambient electron and ion currents at potential ϕ. The physical meaning of the exponential term, also called Boltzmann term, is repulsion to the ambient electrons, whereas the term with an exponent α gives attraction to the incoming ions. The secondary and backscattered electrons are accounted by the $<\delta + \eta>$ term, as explained in Chapter 5.

For a positively charged spacecraft ($\phi > 0$), a current balance equation, as controlled by the ambient electron and ion currents, can be formulated. A good approximation is as follows:

$$I_e(0)\left(1 - \frac{q_e \phi}{k\,T_e}\right)^{\alpha} - I_i(0)\exp\left(-\frac{q_i \phi}{k\,T_i}\right) = 0 \qquad (12.2)$$

In (12.2), the secondary electron current is neglected because the positive spacecraft potential attracts the low energy electrons back. The backscattered electron current is always small.

If photoelectron current I_{ph} in sunlight is emitted, a conducting spacecraft likely charges to a positive potential, which is typically a few volts. The current balance equation, (12.2), becomes

$$I_e(0) - I_i(0)\exp\left(-\frac{q_i \phi}{k\,T_i}\right) = I_{ph} \qquad (12.3)$$

In (12.3), the Mott-Smith and Langmuir attraction term, which features an exponent α, is neglected because the positive potential ϕ, induced by photoelectron emission, is often much less than the ambient electron temperature T_e.

12.2 BEAM EMISSION

For beam emissions, the beam current is included in the current balance equation. If the beam current I_B is much smaller than that of the incoming ambient electrons, the spacecraft potential remains negative. (12.1) becomes as follows.

$$I_e(0)\left[1 - \langle \delta + \eta \rangle\right]\exp\left(-\frac{q_e \phi}{kT_e}\right) - I_i(0)\left(1 - \frac{q_i \phi}{kT_i}\right)^{\alpha} = I_B \qquad (12.4)$$

where ϕ is negative.

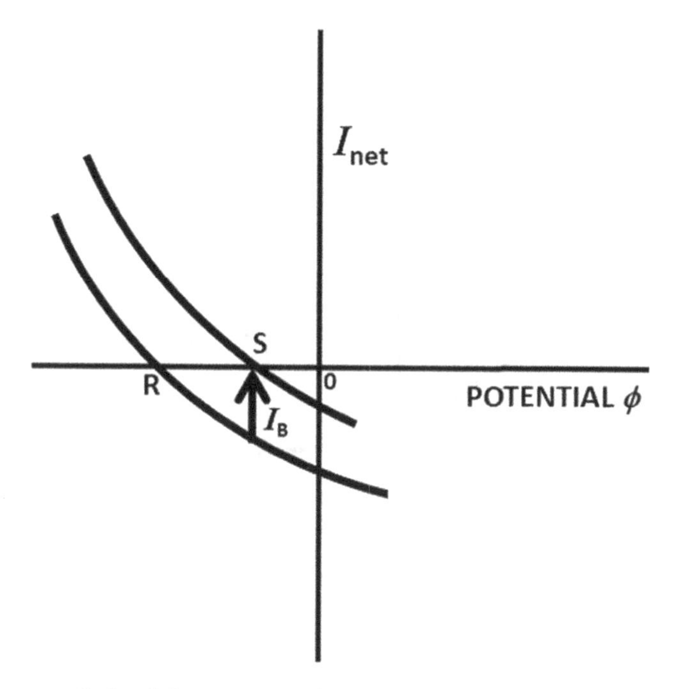

FIGURE 12.1 Shift of the current-voltage curve. The root R is at the crossing of the current-voltage curve. With beam emission I_B, the curve is shifted with the new crossing (root) at S.

One can solve a current balance equation, such as eqs(12.1–12.3), numerically or graphically. To visualize its graphical solution, one way is to seek the solution of a current-voltage curve as follows. Define a net current I_{net} and plot it as a function of the spacecraft potential. The zero-crossing is the potential that satisfies the current balance [Figure 12.1].

$$I_{net} = I_i(0)\left(1 - \frac{q_i \phi}{k T_i}\right)^{\alpha} - I_e(0)\left[1 - \langle \delta + \eta \rangle\right]\exp\left(-\frac{q_e \phi}{k T_e}\right) \quad (12.5)$$

In Figure 12.1, if the potential ϕ increases to large values negatively, the effective current increases positively. This means more positive ions would be attracted to the spacecraft.

If an electron beam current I_B is emitted, the entire current-voltage curve is shifted upward by the amount I_B [Figure 12.2]. Mathematically, this is caused by the additional term I_B to the effective current.

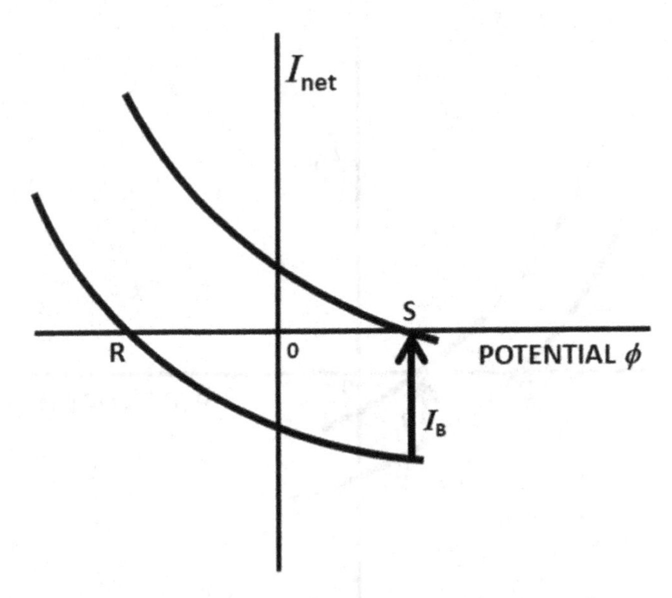

FIGURE 12.2 Shift of the current-voltage curve. With a larger beam current I_B, the root is shifted to S at a positive potential.

$$I_{eff} = I_i(0)\left(1 - \frac{q_i\phi}{k\,T_i}\right)^{\alpha} - I_e(0)\left[1 - \langle\delta + \eta\rangle\right]\exp\left(-\frac{q_e\phi}{kT_e}\right) + I_B \quad (12.6)$$

The solution ϕ is given by the zero-crossing of the curve. At the crossing, I_{net} is zero, meaning that the current balance, (12.4), is satisfied. This is why the value of ϕ at the crossing is the solution.

If the electron beam current exceeds that of the ambient electrons, the spacecraft potential becomes positive. (12.2) becomes as follows:

$$I_e(0)\left(1 - \frac{q_e\phi}{kT_e}\right)^{\alpha} - I_i(0)\exp\left(-\frac{q_i\phi}{kT_i}\right) = I_B \quad (12.7)$$

where ϕ is positive.

For positive ion emission, one needs to change the charge signs appropriately.

12.3 BEAM RETURN

As the beam takes electrons away from a spacecraft, the spacecraft potential responds to the beam emission. The exact value of the

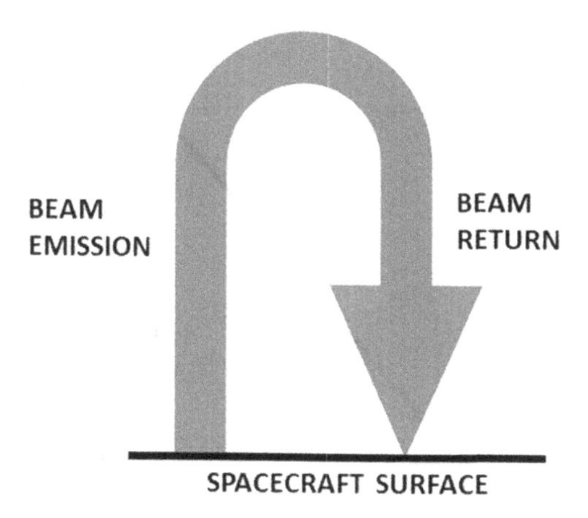

FIGURE 12.3 Beam emission and beam return.

spacecraft is governed by the balance of currents. As the beam current increases, the spacecraft potential increases accordingly. At a sufficiently high beam current, the spacecraft potential energy reaches a critical value that equals the beam energy E_B. Any further increase in the beam current would result in beam return [Figure 12.3]. The net beam current I_{net} emitted is given as follows:

$$I_{net}(\phi) = I_B - I_r(\phi)\Theta(\phi - \phi_C) \tag{12.8}$$

where the step function is zero if the spacecraft potential ϕ is less than the maximum potential ϕ_c, which equals the potential of the beam energy E_B.

This is analogous to throwing a rock upward. It can only reach a critical height at which the rock's potential energy equals its initial kinetic energy [Figure 12.4].

As another analogy, when using a spade to dig a hole on the ground, the hole deepens as more and more soil is dug away. There is a limit to the depth. According to a popular saying: "You can dig as deep as you can throw" [Figure 12.5].

If the beam energy is not single valued, it may form an energy distribution from E_1 to E_2. The portion of the beam with energy E satisfying $E_1 < E < e\phi$ returns. Here, ϕ is the spacecraft potential. The other portion with E satisfying $e\phi(x) < E < E_2$ escapes [Figure 12.6].

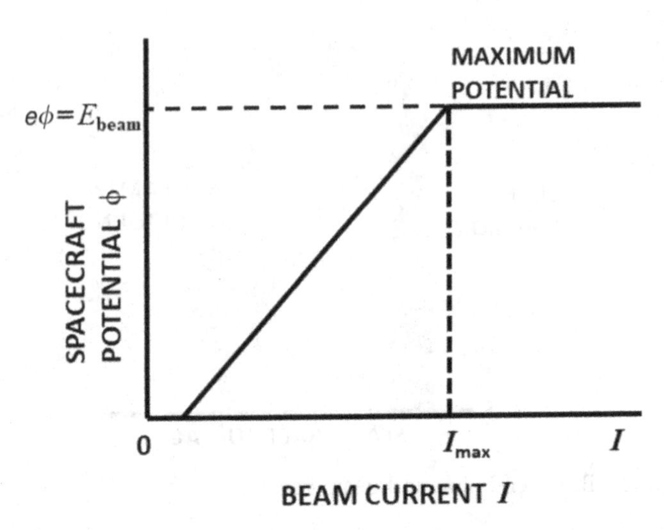

FIGURE 12.4 Maximum potential induced by beam emission. Beyond the critical current I_{max}, any further increase in beam current cannot raise the potential.

FIGURE 12.5 "You can dig as deep as you can throw." This is analogous to trying to increase the spacecraft potential by beam emission. The maximum spacecraft potential equals the beam energy, but not more.

12.4 SUPERCHARGING

The principle "You can dig as deep as you can throw" is applicable only to a closed system without outside interference. Such a system is similar to one describable by a Hamiltonian in quantum mechanics or statistical mechanics. If there is outside interference, the Hamiltonian has to include the outside system responsible for the interference.

As an example, in the situation in Figure 12.5, if someone catches the soil from above and takes it away, the digger can dig deeper than

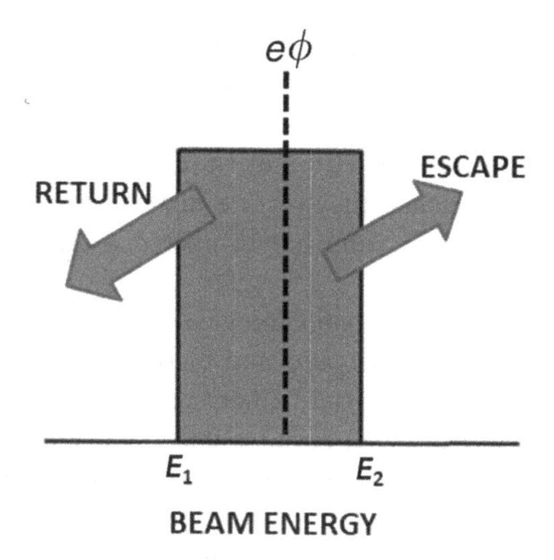

FIGURE 12.6 Beam return and escape from a beam with an energy spread. The portion of the beam with energy E less than the spacecraft potential energy $e\phi(x)$ at position x returns. The other portion with $E > e\phi(x)$ escapes.

without the person catching soil. But, then, the Hamiltonian has to include the system of the catcher. As another example, a spacecraft travels at a speed so fast that, by the time the beam particles return, the spacecraft has already moved away. In any case, if beam emission can charge a spacecraft to a potential higher than that of the beam energy, the phenomenon is called supercharging. In a closed system, supercharging does not occur [*Lai*, 2002].

One common method for measuring spacecraft potential is to use booms. It is assumed that the tip of the boom is outside the sheath potential of the spacecraft body. That is, it is assumed that the tip of the boom is at zero potential. With this assumption, the potential difference between the potential of the spacecraft body and that of the tip would indicate the spacecraft potential. For example, as a result of high-current electron beam emission, the spacecraft body may charge to a positive potential, which equals the beam energy. However, the tip of the boom may charge to a negative potential because the tip intercepts abundant incoming electrons. Thus, the potential difference may exceed the true spacecraft body potential. One has always to be careful in providing physical interpretation to the data obtained in experiments.

A necessary comment on the charging of the boom tip is offered as follows. High-current alone does not charge a spacecraft surface [see Chapters 6 and 11]. To charge a surface, the current has to have abundant electrons that are hot enough. If the hot electrons are from the ambient plasma, they form a distribution with a temperature T. If so, the temperature has to exceed a critical value T_c for charging to occur.

In the high-current beam emission case, however, the beam energy E_B is likely nearly mono-energetic, and therefore the return current is likely of energy E_B approximately. Charging by a near mono-energetic electron current of energy E_B requires that the energy E_B exceeds the second crossing point E_2 of the secondary yield $\delta(E)$ of the surface material [see Section 5.2].

For example, if E_2 of the surface material is 10 keV, then mono-energetic electrons with kinetic energy exceeding 10 keV would charge the surface upon impact. If the impacting charges are a mixture of electrons and ions of various energies, then one has to integrate over the distributions for their currents. The currents are to be used in a current balance equation for determining the onset of charging or the resulting potential of the surface.

12.5 THE DRIVING FORCE AND THE RESPONSE

The spacecraft potential is governed by the balance of currents. If one of the currents varies significantly, the spacecraft potential varies accordingly. One can consider the significantly varying current as the driving factor, and the spacecraft potential as the response. Since the beam current is artificially controlled from the ground, one can vary it methodically, observe the spacecraft potential in response, perform the data analysis of the results obtained, and provide meaningful physical interpretations. In this manner, the reader can design many experiments of his/her own in the future for studying the response to the driving force in various manners, provided that the currents of nature do not change rapidly during the time span of the experiment.

As examples, two electron beam experiments performed on the SCATHA satellite feature electron beam current variation and the spacecraft potential in response [*Lai, et al.*, 1987; *Lai*, 1994]. For more detailed perspective, interested readers can study the data analysis in these referenced examples.

12.6 BEAM DIVERGENCE

This section derives the electric field of a cylindrical electron or ion beam. From the Gauss law, the surface integral of the electric field E is related to the volume integral of the charge density ρ [Figure 12.7].

$$\iint d\mathbf{S}.\,\mathbf{E} = \iiint dV\frac{\rho}{\varepsilon_0} \tag{12.9}$$

For a long cylinder, the Gauss law yields

$$E = \frac{r\rho}{2\varepsilon_0} \tag{12.10}$$

where r is the radius of the cylinder. The current I is given by the flux ρv multiplied by the cross-section πr^2. The charge velocity is denoted by v.

$$I = \pi r^2 \rho v \tag{12.11}$$

From Biot-Savart's law, the induced magnetic field B is given by

$$B = (\mu_0/2\pi)I/r \tag{12.12}$$

From Faraday's law, the induced electric field E' is given by

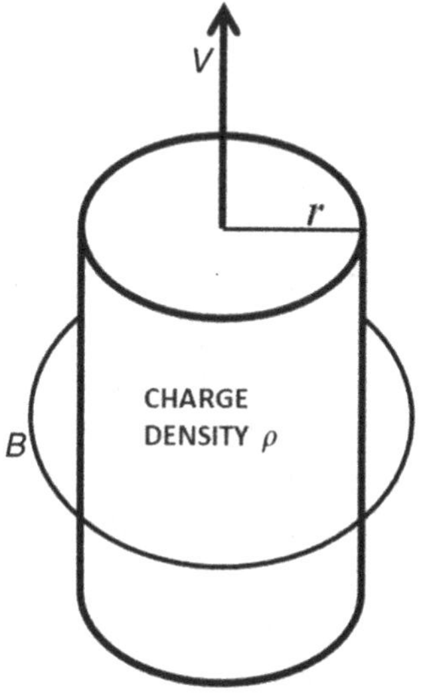

FIGURE 12.7 A cylindrical charged particle beam.

$$E' = v \times B = (\mu_0/2\pi) vI/r \tag{12.13}$$

The net outward force on a beam electron at r is given by

$$eE_{\text{net}} = eE - eE' = \frac{eI}{2\pi\varepsilon_0 rv}\left(1 - \frac{v^2}{c^2}\right) \tag{12.14}$$

The equation of motion of the electron is given by

$$m\frac{d^2r}{dt^2} = \frac{1}{4\pi\varepsilon_0}\frac{2eI}{rv}\left(1 - \frac{v^2}{c^2}\right) \tag{12.15}$$

The divergence rate $(dr/dt) = 0$ for $r = r_0$ at $t = 0$.

(12.15) can be integrated to give the electron trajectory along the expanding envelope of the beam.

For a spacecraft charged to an attractive potential, ϕ, the velocity v slows down as the beam electron travels outward. The velocity $v(x)$ depends on the spacecraft potential profile $\phi(x)$, where x is the distance from the spacecraft surface. For simplicity, the Debye form is often useful [*Whipple, et al.*, 1974].

$$\phi(x) = \frac{\phi_S R}{x + R}\exp\left(-\frac{x}{\lambda}\right) \tag{12.16}$$

where ϕ_s is the surface potential, R the spacecraft radius, and λ the Debye distance.

12.7 BEAM EMISSION FOR SPACE PROPULSION

Ion beam emission for spacecraft propulsion has been of interest for decades. Traditionally, the ion beam devices feature an ionization chamber, into which neutral gas is injected and ionized. Upon ionization, pair creation of positive ions and electrons are created. The ions are extracted by a grid where a high negative voltage is applied. The ionization rate is often low, 6% for example. The positive ions are extracted with typically keV energies. The electrons are repelled by the grid and cannot escape. Some neutral gas, which is uncharged, can wander out. Charge exchange between the keV ions and the slow neutrals may generate some slow ions and high-energy neutrals.

For prevention of spacecraft charging, the ion beam is neutralized by means of electron injection into the beam just outside the beam exit point. This has been the general idea used in conventional ion beam devices in recent decades. There are many variations, each with advantages and disadvantages. For a comprehensive review of various ion propulsion devices, see *Martinez-Sanchez and Pollard* [1998].

In recent years, the use of ionic liquid ion source has been of interest for future space propulsion [*Lozano and Martinez-Sanchez*, 2005; *Coffman, et al.*, 2019]. Some examples of negative and positive ionic liquid ions are BF_4^- and EMI^+, respectively. The ionic liquid can evaporate from a sharp point, which is called a Taylor cone. The criterion for evaporation is that the electric field at the Taylor cone has to exceed the surface tension of the liquid.

$$\gamma \nabla . \hat{n} = \frac{1}{2} \varepsilon_0 E^2 \qquad (12.17)$$

where $\hat{n}$ is the surface normal unit vector, $\nabla . \hat{n}$ the curvature of the ionic liquid meniscus, γ the surface tension, and E the electric field at the cone tip. This method enables very small and very light ion beam devices to be built. The beam ions can travel at multiple keV energy. Conceivably, smaller and smaller devices can be developed in this approach, and, if so, they may be able to propel even small and unidentifiable flying objects to high speeds.

Ion beam emissions from a spacecraft take away ions from it, resulting in charging of the spacecraft. With ionic liquid ions, one can emit two beams, one being positive and one negative, in a bimodal manner. If the opposite currents are equal, no net charge is emitted. The result is no charging induced by the beam emission. As a result, there is no need for mitigation. If the bimodal beams are too near each other [Figure 12.8], one may need to consider their mutual attraction by their net electric fields (eq.12.12).

In a possible scenario, if the spacecraft is traveling in a region where the space plasma is hot during stormy space weather, natural charging to high negative potential may occur. In that case, one beam returns while the other beam continues to carry out the task of propulsion [Figure 12.9].

12.8 SPACECRAFT DAMAGE BY BEAM EMISSION

As an example of non-environmental cause for spacecraft anomalies, we cite the case of artificial beam emission from the SCATHA (P78–2) satellite [*Cohen, et al.*, 1981]. A 3 keV 13 mA electron beam was emitted

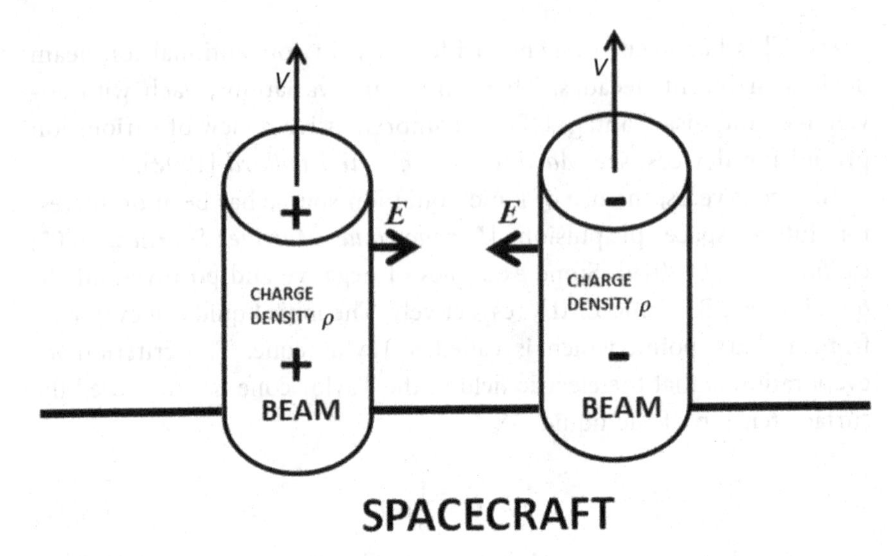

FIGURE 12.8 Electric field attraction between a positive and a negative cylindrical beam.

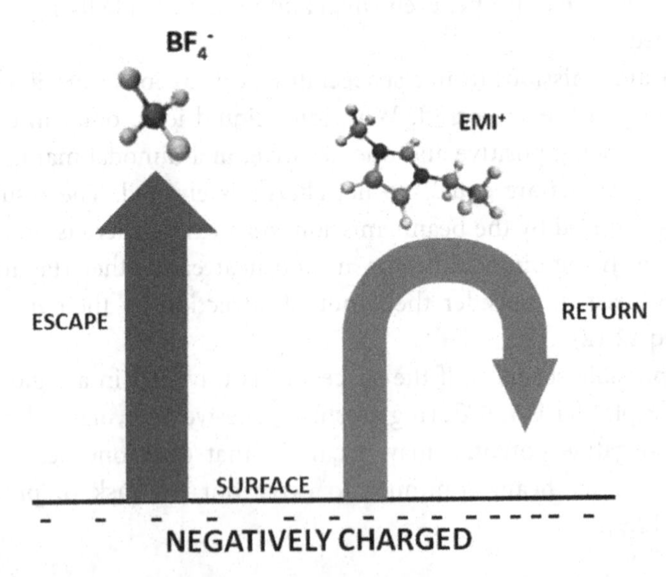

FIGURE 12.9 Emission of a positive and a negative ion beam from a spacecraft. If the spacecraft charge to high negative voltages in stormy space weather, one of the beams returns.

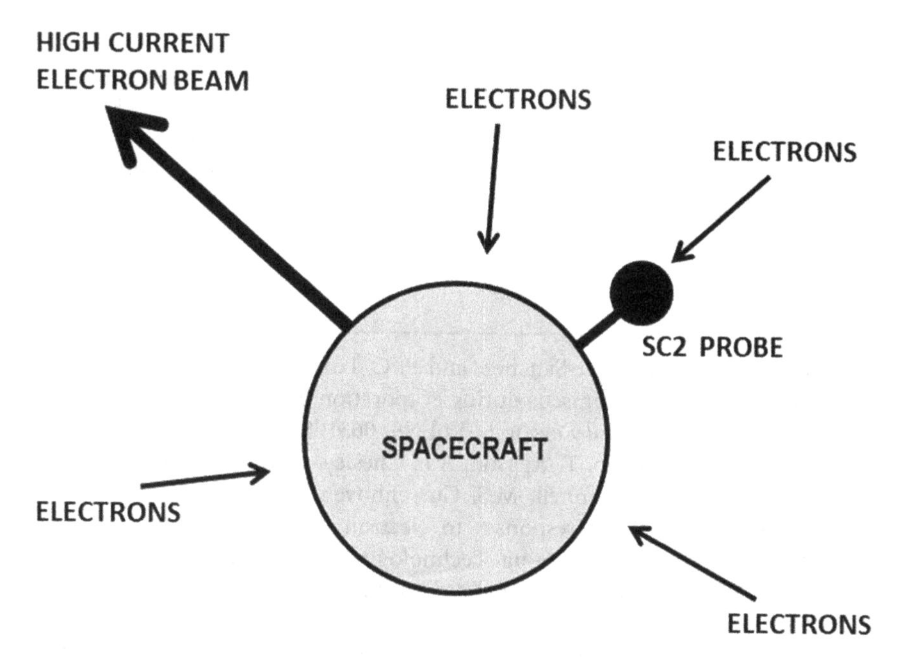

FIGURE 12.10 Spacecraft electronics knocked out by beam emission.

from SCATHA. It promptly knocked out an instrument (SC2 probe) on the surface of the spacecraft even though the instrument was not aimed at by the beam. The cause and effect were clearly identified in this example. In this type of anomaly, the damage can be total and permanent [Figure 12.10].

EXERCISES

1. The spacecraft potential at equilibrium is governed by current balance. Consider a cylindrical spacecraft emitting an ion beam current I_b. Write down the current balance equation.

   Suppose the ion beam emission induces a negative spacecraft potential ($\phi < 0$) so high ($-q_i\phi >> kT_i$ and $-q_e\phi >> kT_e$) that all ambient electrons are repelled. The exponential term for the electrons is negligibly small. Show that the spacecraft potential can be written as follows:

$$\phi = -\frac{kT_i}{q_i}\left(\frac{I_b}{I_i(0)}\right)^2$$

2. Suppose a positive ion beam of energy 200 eV is emitted from a spacecraft that is charged to –6000 V. Assuming that the ion beam current is large enough to generate substantial ion-induced secondary electrons, what is the final potential of the spacecraft?

REFERENCES

Coffman, C.S., M. Martínez-Sanchez, and P. C. Lozano, Electrohydrodynamics of an ionic liquid meniscus during evaporation of ions in a regime of high electric field, *Physical Review E*, Vol. 99, 063108, (2019).

Cohen, H.A., R.C. Adamo, T. Aggson, A.L. Chesley, D.M. Clark, S.A. Damron, D.E. Delorey, J.F. Fennell, M.S. Gussenhoven, F.A. Hanser, et al. P78-2 satellite and payload responses to electron beam operations on March 301979. Spacecraft Charging Technology – 1980. NASA-CP-2182, Oct 1981, N.J. Stevens, C.P. Pike (Eds.), *ADA114426*, NASA Lewis Research Center, Cleveland, OH, 509–559, (1981).

Lai, S.T., An improved Langmuir probe formula for modeling satellite interactions with near geostationary environment, *J. Geophys. Res.*, Vol. 99, 459–468, (1994).

Lai, S.T., On supercharging; Electrostatic aspects, *J. Geophys. Res.*, Vol. 107 (A4), (2002), 10.1029-10.1038.

Lai, S.T., H.A. Cohen, T.L. Aggson, and W.J. McNeil, The effect of photoelectrons on boom-satellite potential differences during electron beam ejections, *J. Geophys. Res.*, Vol. 92 (A11), 12319–12325, (1987).

Lozano, P., and M. Martinez-Sanchez, Ionic liquid ion sources: Characterization of externally wetted emitters, *J, Colloid and Interface Sci.*, Vol. 282 (2), 451–421, (2005).

Martinez-Sanchez M., and J.E. Pollard, Spacecraft electric propulsion – An overview, *J. Propulsion and Power*, Vol. 14 (5), (1998). 10.2514/2.5331.

Whipple, E.C., Jr., J.M. Warnock, and R.H. Winckler, Effect of satellite potential in direct ion measurements through the magnetopause, *J. Geophys. Res.*, Vol. 79, 179–186, (1974).

CHAPTER 13

Mitigation Methods

13.1 ACTIVE AND PASSIVE METHODS

Spacecraft charging to high negative voltage may interfere with the scientific measurements onboard and harm the electronic instruments. It may also affect the spacecraft navigation or even terminate the spacecraft mission. Therefore, it is necessary to mitigate the high voltage caused by spacecraft charging.

Any low-level voltage due to charging can be tolerated usually, because it does not cause significant interference in the electronics, nor does it cause severe harm on the instruments.

In general, there are active and passive methods for mitigation of spacecraft charging. In an active method, a command is needed to activate the mitigation process. In a passive method, a command is not needed. Table 13.1 lists active and passive methods that are currently known.

13.2 FIELD EMISSION OF ELECTRONS

If a spacecraft charges to a high negative potential with excess electrons, an obvious method of mitigation is to throw away the electrons from the spacecraft. When the electric field E is sufficiently high at a sharp point, field emission [*Fowler and Nordheim*, 1928] occurs at the sharp point. The excess electrons can be ejected automatically in this way at the sharp points installed on a spacecraft. An advantage of this method is that it does not require a command to be sent from the ground to the spacecraft.

DOI: 10.1201/9780429275043-13

TABLE 13.1 Active and Passive Mitigation Methods

Method	Physics	Type	Comment
Sharp Spike	Field Emission	Passive	Requires High E Field at the Sharp Spike. Ion Sputtering of the Sharp Points. Mitigates Charging of Conducting Ground Surface But Not Dielectrics. Differential Charging Ensues.
Conducting Grids	Prevent High Differential Potentials	Passive	Periodic Surface Potential.
Semi-Conducting Paint	Increased Conductivity on Dielectric Surfaces	Passive	Mitigates Dielectric Surface Charging. Paint Conductivity May Change Gradually.
High Secondary Electron Yield Material	Secondary Electron Emission	Passive	Ambient Electrons with Energies Above the Second $\delta(E) = 1$ Crossing Point Can Cause Charging.
Hot Filament	Thermal Electron Emission	Active	Space Charge Current Limitation. Mitigates Conducting Ground Charging Only. Differential Charging Ensues.
Electron Beam	Emission of Electrons	Active	Mitigates Conducting Ground Charging Only. Differential Charging Ensues.
Ion Beam	Return of Low-Energy Ions	Active	Neutralizes the "Hot" Spots. Effective for Both Conducting and Dielectric Surfaces. The Ions May Act As Secondary Electron Generators. Cannot Reduce the Potential Below the Emitted Ion Energy.
Plasma Emission	Emission of Electrons and Ions	Active	More Effective than Electron or Ion Emission Alone.
Evaporation	Evaporation of Polar Molecules which Attach Electrons	Active	Mitigates Conducting and Dielectric Surface Charging. Not Intended for Deep Dielectric Charging. May Cause Contamination.

(*Continued*)

TABLE 13.1 (Continued)

Method	Physics	Type	Comment
Metal-Based Dielectrics	Increased Conductivity in Dielectrics	Passive	Mitigates Deep Dielectric Charging. Metal-Based Material Needs to Be Homogeneous to Be Useful. Conductivity Change and Control Need to Be Studied.
Mirrors or LED	Photoemission from the Shadowed Side	Active	Mitigation of the High-Level Charging on the Shadowed Side. Removal of the Photoelectron Barrier on the Sunlit Side.

The physical principle of field emission requires quantum mechanics and is outside the scope here. An exact calculation, which we do not need here, would have to input the Fermi energy μ of the material [von Engel, 1983].

To get some idea, let us give a simple example. A sharp point of a typical material with work function W = 5 V and sharp point radius = 0.5 nm requires a critical electric field E of about 10^{10} V m^{-1} for field emission of electrons [*von Engel*, 1983].

Positive ions are also emitted from sharp points by positive electric fields about ten times stronger. As a side remark, modern space propulsion of very small satellites typically uses charged liquid metals, positively or negatively charged, emitting from very sharp points at kilovolts [*Krejci and Lozano*, 2018; *Lai and Miller*, 2020].

The thermionic emission method can also be used for mitigation of spacecraft charging by means of electron emission. This method is similar to the field emission method, but it uses a hot filament for emitting electrons from the spacecraft surface. The mathematics of thermionic emission is described by modified Richardson's equation:

$$J = \lambda A_0 T^2 \exp(-W/kT) \tag{13.1}$$

where J is the electron flux, T the filament temperature in °K, W the work function of the filament material, k the Boltzmann constant, λ a material-dependent parameter, and A_0 a constant given by $A_0 = 1.20173 \times 10^6\ Am^{-2}K^{-2}$.

The thermionic emission method is an active method. It requires a command from the ground to turn on the hot filament. Hot filaments consume significant electric power. Also, prolonged usage burns out the filament.

13.3 DISADVANTAGE OF BOTH METHODS USING ELECTRON EMISSION

If the spacecraft is covered by all-conducting surfaces, the electron emission methods are applicable because the excess electrons would be emitted away. However, if the spacecraft is covered with conducting and dielectric surfaces, these methods are not applicable. Applying this method may leave the surfaces in a differential charging situation, which is worse than uniform charging. Although the excess electrons emitted can be drawn from the conducting surfaces connected to the electron emitter, the electrons in the dielectric surfaces cannot. This is because dielectric surfaces are nonconductors, which hardly allow electrons or ions to travel through them [Figure 13.1].

For example, suppose the surfaces are bombarded by hot electrons of multiple keV in energy. As a result, all the surfaces charge to negative kilovolts. Let us say that the potentials are approximately −2 kV, for example. Suppose, then, electrons are emitted from the emitters for mitigation purpose. The conductor surface potentials would drop to nearly zero volt as a result, but the dielectric surfaces are untouched

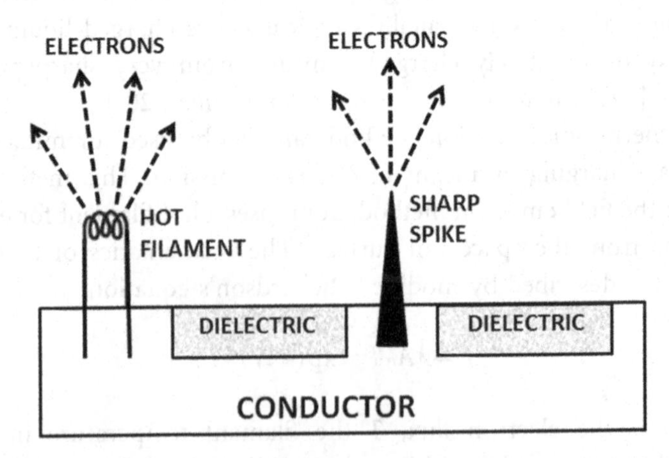

FIGURE 13.1 Mitigation method by emitting electrons. This method works for spacecraft with all-conducting surfaces but not for spacecraft covered with conducting and dielectric surfaces.

because electrons cannot travel through dielectric materials, or almost so. Therefore, the conducting surface potentials are about zero volt, while the dielectric surface potentials are about −2 kV. Such a situation is differential charging, which is more dangerous than the original one.

13.4 LOW-ENERGY ION EMISSION

Low-energy positive ion emission from a highly negatively charged spacecraft can produce an ion fountain. The returning ions can find their way to the most negatively charged spots on the spacecraft, thereby neutralizing the negative charge. The low-energy ion beam emission method is effective in mitigating differential charging [*Lai*, 2003; *Lai and Cahoy*, 2017]. For a simple numerical model of trajectories of charged particles emitted from, and returning to, attractively charged surfaces, see, for example, *Carini et al.* [1986].

If positive ions with more energy than the spacecraft potential energy are being emitted, they would leave. As a result, the spacecraft potential increases with negative voltage until the potential energy equals the ion beam energy. The negative potential is clamped by the ion beam emission. However, if there are thermal neutrals present, the potential level can decrease further because charge exchange between the emitted ions and the thermal neutrals can occur. As a result, the newborn positive ions, being of nearly the same energy as the neutrals, would return and participate in the mitigation process.

13.5 LOW-ENERGY PLASMA EMISSION

Low-energy missions of electrons and positive ions are more efficient than low-energy ion emission alone [*Lai*, 2003]. While electron emission takes away the surface charging electrons from the conductors, the returning positive ions help neutralize the dielectric surfaces.

In the DSCS experiment [*Lai*, 2003], it was demonstrated that as soon as low-energy plasma (as ionized neutral gas) was emitted, the preexisting differential charging between kapton (red) and quartz (green) surfaces immediately disappeared. However, this method requires an active command. Soon after the duration of plasma emission in the DSCS experiment [Figure 13.2], differential charging appeared again (at 10,000 UT). In Figure 13.2, the zero level had its own offset problem.

In practice, there is no need to emit a high-density plasma beam, because the ambient electron flux, 0.115×10^{-9} Acm^{-2}, responsible for charging is low. The plasma emitted should not be of high energy, otherwise it would not return.

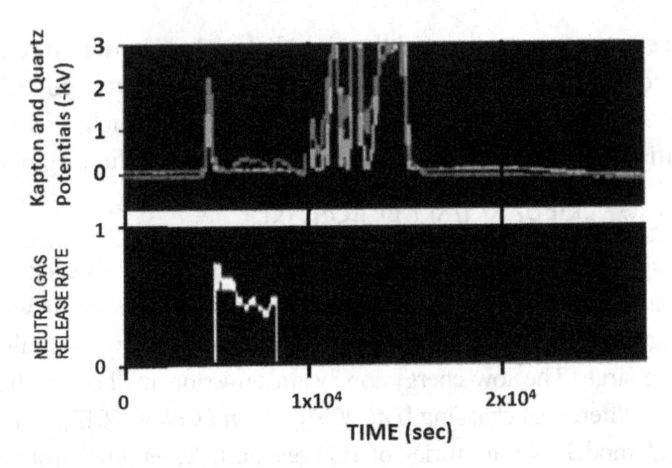

FIGURE 13.2 Demonstration of charging mitigation by using low-energy plasma release. The plasma from the ionized neutral gas reduced the charging potentials of the kapton (red) and quartz (green) potentials. The zero level had its own offset problem.

This method, though efficient, is not practical, not only because it requires a command from the ground to activate it, but also it requires a supply of neutral gas for making plasma. Both the neutral gas supply and the period of mitigation are finite.

13.6 PARTIALLY CONDUCTING PAINT

In spacecraft designs for certain mission objectives, the need for highly insulating surfaces and highly conducting surfaces may be relaxed. If so, partially conducting paint, such as indium oxide, may be used for reducing differential charging of spacecraft surfaces. The paint renders the surfaces conducting with each other to some extent. Since no excess electrons are expelled, the absolute potential relative to the ambient plasma is unchanged. Absolute, but uniform, charging to high-level can still occur, but differential charging can be avoided under most conditions.

If the partially conducting surface property is acceptable, this mitigation method is effective and convenient. It automatically prevents differential charging of the spacecraft surfaces, without the many steps required in the methods of electron or plasma emissions.

13.7 SPRAY OF POLAR MOLECULES

Spray of polar molecules, such as water, on conducting and nonconducting spacecraft surfaces can also get rid of the surface electrons. It does so by means of scavenging the surface electrons, thereby accumulating the electrons in the polar molecule droplets.

When the droplet Coulomb repulsion force exceeds the surface tension, the droplet bursts into smaller ones. The smaller droplets have shorter radii, and therefore they have larger Coulomb repulsion force and burst into even smaller ones. In this manner, the charged droplets evaporate away [Lai and Murad, 1995, 2002a, 2002b]. Unlike the electron emission methods, this method gets rid of electrons from conducting and insulating surfaces alike.

A disadvantage of this method is that it is an active one, requiring a command to commence. It needs a finite supply of polar chemicals, and it may contaminate surfaces.

13.8 MITIGATION BY USING MIRRORS

Spacecraft with mostly dielectric surfaces can charge in sunlight because the dark side can charge to negative volts without emission of photoelectrons (see Chapter 10). The high-negative-voltage contours from the dark side can wrap around to the sunlit side, forming a barrier and blocking the photoelectrons [*Besse and Rubin*, 1980; *Lai and Tautz*, 2006]. Low-energy electrons, such as photoelectrons and secondary electrons, can be trapped behind the barrier on the sunlit side [*Lai and Cahoy*, 2015].

Sunlight reflected by mirrors can generate photoelectrons from the dark side, thereby mitigating the surface charging [Figure 13.3]. Aluminum

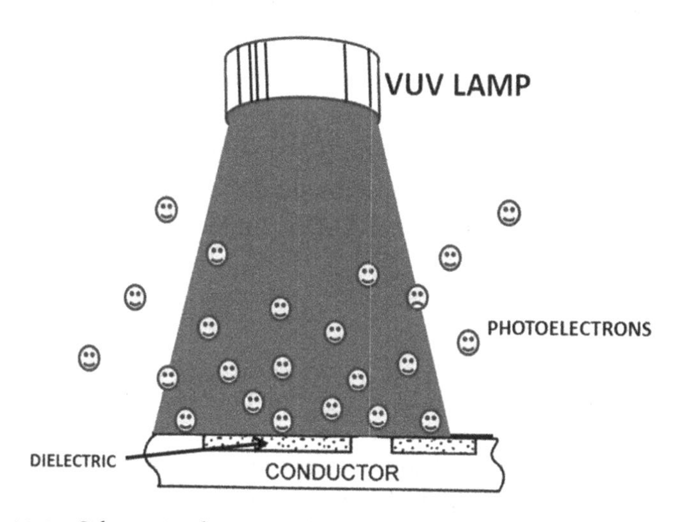

FIGURE 13.3 Schematic diagram of using ultraviolet light for photoemission from surfaces. Removing the surface electrons reduced the charging level of conductor and dielectric surfaces in eclipse and in sunlight.

mirrors can have reflectivity around 88% for ultraviolet light. The mirrors have to be deployed at a distance from the spacecraft and, like solar panels, have to face the sun at an appropriate angle for reflection [*Lai*, 2012].

For satellites with all-conducting surfaces, charging is usually not a problem because the photoelectron flux emitted from the spacecraft surfaces exceeds the ambient electron flux by about two orders of magnitude typically. However, some satellite designs feature shadowed areas much larger than sunlit areas or even shadows in shadows, resulting in a larger ratio of ambient electron current to that of photoelectrons. Also, during extreme geomagnetic storms, the ambient electron flux may surge to defeat the photoelectron flux. In such situations, the mirrors reflecting sunlight to the shadowed surfaces would help for mitigation of spacecraft charging of even conducting spacecraft in sunlight in extreme space weather.

This method [*Lai*, 2012] should be effective. It is an opportunity for the space community to test this method in the laboratory and in space.

Unlike the plasma emission method, there is no need to operate a plasma discharge chamber attached to a finite gas supply. There may be need for AI (artificial intelligence) to control automatically the tilting of the mirrors in accordance with the sun angles.

13.9 MITIGATION BY USING LED

Light emitting diodes (LED) manufactured years ago used to emit low-frequency light only. Modern advances [*Nobel prize*, 2014] have been successfully extending the frequency range toward the violet side. The method of using LEDs to generate photoelectrons from a spacecraft works similarly to the mirror reflection method, but the LED method works at all times, even in eclipse. A disadvantage is that it requires a power supply, and the lifetime of LED is finite.

EXERCISES

1. What is the space charge electric field in the radial direction of the side of a cylindrical charged particle beam? One can use a parameter λ (0 to 1) to indicate the ratio of electron to ion densities in the beam. If the cylindrical charged particle beam is emitted from the spacecraft surface, the electric field can compete with that of the spacecraft surface, which is charged to a potential. [Hint: Eq (8) in *Lai*, 2002].

2. What are the weaknesses of using sharp spikes for mitigating negative voltage charging?
   a. Electrons can be drawn from the conductors only and not from the dielectrics.
   b. Ambient ions attracted by the high electric field of the sharp tip can blunt the tip.
   c. Sharp spikes can hurt fingers during assembly.
   d. All of the above.
3. What is the physical mechanism that enables plasma emission to be a very effective method in mitigating spacecraft charging?
4. Positive ions of energy 200 eV are emitted from a spacecraft that is charged a prioi to −6000 V. Assume that the returning ion current is large enough to generate substantial ion-induced secondary electrons, even though the ion-induced secondary electron yield $\delta_i(E)$ is small. What is the final potential of the spacecraft [Figure 13.4]?

Solution:
The ion "emission and return" acts as a secondary electron generator. The electrons are repelled by the negative potential of the satellite. As a result, the magnitude of the final potential is −200 V.

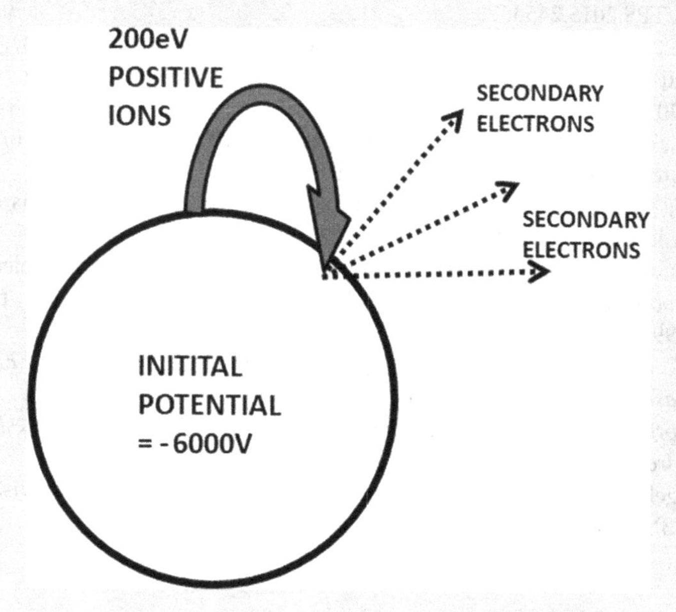

FIGURE 13.4 Mitigating a spacecraft charged a priori to −6000 V.

REFERENCES

Besse, A.L., and A.G. Rubin, A simple analysis of spacecraft charging involving blocked photoelectron currents, *J. Geophys. Res.*, Vol. 85 (A5), 2324–2328, (1980).

Carini, P., G. Kalman, Y. Shima, S.T. Lai, and H.A. Cohen, Particle trajectory between charged plates, *J. Phys. – Applied Phys.*, Vol. 19, 2225–2235, (1986).

Fowler, R.H., and L.W. Nordheim, Electron emission in intense electric fields, *Proc. Roy. Soc.*, Vol. A119, 173–181, (1928).

Krejci, D., and P. Lozano, Space propulsion technology for small spacecraft, *Proc. IEEE,* Vol. 106, 362 (2018).

Lai, S.T., A critical overview on spacecraft charging mitigation methods, *IEEE Trans. Plasma Sci.*, Vol. 31 (6), 1118–1124, (2003).

Lai, S.T., On supercharging; Electrostatic aspects, *J. Geophys. Res.*, Vol. 107 (A4), (2002), 10.1029-10.1038.

Lai, S.T., Some novel ideas of spacecraft charging mitigation, *IEEE Trans. Plasma Sci.*, Vol. 40 (2), 402–409, (2012). 10.1109/TPS.2011.2176755. http://ieeexplore.ieee.org/xpl/articleDetails.jsp?arnumber=6118330

Lai, S.T., and K. Cahoy, Spacecraft charging, in *Encyclopedia of Plasma Technology*, J. Shohet (ed.), Taylor and Francis Publishers, United Kingdom, pp. 1352–1366, (2017). 10.101081/E-EPLT-120053644. https://www.taylorfrancis.com/books/e/9781482214314/chapters/10.1081%2FE-EPLT-120053644

Lai, S.T., and K. Cahoy, Trapped photoelectrons during spacecraft charging in sunlight, *IEEE Trans. Plasma Sci.*, Vol. 43 (9), 2856–2860, (2015), 10.1109/TPS.2015.2453370.

Lai, S.T., and C. Miller, Retarding potential analyzer: Principles, designs, and space applications, *AIP Advances*, Vol. 10, 095324, (2020), 10.1063/5.0014266.

Lai, S.T., and E. Murad, Mitigation of spacecraft charging by means of ionized water, U.S. Patent No. 6,463,672 B1, (2002a).

Lai, S.T., and E. Murad, Mitigation of spacecraft charging by means of polar molecules, U.S. Patent No. 6,500,275 B1, (2002b).

Lai, S.T., and E. Murad, Spacecraft charging using vapor of polar molecules, in *Proceedings of the Plasmadynamics Lasers Conf.*, AIAA-95-1941, pp. 1–11 (1995).

Lai, S.T., and M. Tautz, Aspects of spacecraft charging in sunlight, *IEEE Trans. Plasma Sci.*, Vol. 34 (5), 2053–2061, (2006).

Nobel prize, (2014). Available at http://www.nobelprize.org/nobel_prizes/physics/laureates/2014/press.html.

von Engel, A., *Electric Plasmas: Their Nature & Uses*, Taylor & Francis, (1983), ISBN 0850661471, 9780850661477.

Index

Printed in the United States
by Baker & Taylor Publisher Services